Practical Physics Labs

A RESOURCE MANUAL

$$mgh = \frac{1}{2}I\omega^2$$

Peter Goodwin

illustrated by Nicholas Soloway

J. Weston Walch, Publisher

Portland, Maine

1 2 3 4 5 6 7 8 9 10

ISBN 0-8251-1683-X

J. Weston Walch, Publisher

P.O. Box 658 • Portland, Maine 04104-0658

Printed in the United States of America

To my young son, Hunt.
May he someday enjoy the study of science.

Contents

LAB #	TITLE	

About This Book

The labs in this book can be conducted with only a small investment in equipment. Most physics labs have power supplies, an air track, and some of the other equipment that is used. Alternative, less expensive materials are often suggested so you can run the labs without having to break your budget.

These labs are directly related to physics principles, and the write-ups direct students toward a better understanding of the concepts. Some of the labs are the kind students will remember when they return for reunions.

Each lab is separated into three sections. The first is the handout which must be duplicated for the students. This section introduces the lab, gives the procedure, guides the students through the analysis of data, and asks questions for the conclusion. An attempt has been made to make each step in the lab short so that they are easy to follow. This increases the number of pages which need to be duplicated, but it helps make the lab run smoothly.

The second section, the Teacher Notes, gives the rationale for the lab and how it applies to what the students are doing. It also helps in the organization of the lab materials and in some cases tells how to build them. This section points out problem areas in running the lab and suggests how to smooth out these rough spots. Finally, the Teacher Notes answer the questions asked in the Conclusions and give the answers to the short quiz on the lab.

The third section is a five-question multiple-choice quiz. This quiz is meant to be given before the students do the lab as a check on whether they have studied the handouts before coming to class. The quiz includes questions on the basics of running the lab and also on the concepts needed for writing up the lab. For instance, if kinetic energy is used in the lab, the quiz will ask questions relating to the calculation of kinetic energy. Therefore, part of the student's preparation for the lab must be reviewing the equations and concepts involved. The quiz does not test knowledge of the lab results—only the preparation for the lab.

Error and How to Report It

With any data gathering, some measurement error exists. Because of this, error may exist in calculated values. Students should understand what error is and how to use it in the lab. On the most basic level, students should be able to say that "Because my measurements are probably a little bit off, my answer probably will be, too." Even this basic knowledge of error allows the student to realize that the experiment can work quite well even though the acceleration due to gravity is calculated as 9.6 meters per second squared rather than 9.8. With no knowledge of error, the 9.6 answer must be wrong and the experiment a failure.

In a further step students would estimate how large their measurement errors are and relate that information to their answers. For instance, if they estimate a 5% error in a value they measure, they might expect that there would be a 5% error in their result. The student must understand percent—most students should be able to apply what they have learned in math class to a physics lab.

A more complete error analysis uses the rules for estimating error: When adding or subtracting, add numerical error. When multiplying or dividing, add percent error. The calculations used to estimate error using these rules are difficult for some students, but the reason for the difficulty is that students do not understand what error means. Error is only an estimate of how far you think you might be off. You are looking for one significant digit, 1%, 5%, 10%, 20%, or 50%, not 1.5454%. By finding a rough estimate of error, students can still get an idea of how far the calculated value might be from the actual, and the calculations can be done quickly. Students should be saying, "1 is about 10% of 9," rather than saying, "1 is 11.1% of 9." They should be looking for "ballpark" figures. For a more complete discussion for students on how to calculate error, see *Physics Can Be Fun: A Sourcebook of Practical Problems* (Peter Goodwin, J. Weston Walch, 1985).

Students must also think about error not related to actual measurements. For example, if something moves down a hill, friction is present to some degree. Is it a large factor, 10% or more? Is it something which can be ignored? These judgments about error are subjective, but they are important for students to make. Sources of error other than measurement error can explain why there is a 20% error even though the estimated measurement error was only 5%. Students should be encouraged to think about the significance of their suggested error sources. With a bicyclist going down a hill, they should realize that friction is probably a larger error factor than the color of the wheels or the bugs which hit the rider in the face. Students must learn to think about which errors are important and which are not.

Finally, sometimes the difference between the answer and the accepted value can't be explained by the estimated errors. Then students should be encouraged to look for mistakes which they have made in data processing. In this way, error calculations help students to use their data correctly.

Graphing

(may be reproduced for your students)

Graphs are used in these labs wherever they help to clarify concepts. Some of the graphs are easy to draw while others are more difficult. All graphs should have appropriate units and scales shown on them. It should be possible to look at the graph and immediately know what has been graphed and how large the graphed values are.

Graphs should not be drawn by connecting the dots. The line should be used to "average the data." Some data points will be above and some will be below the line. This should be true throughout the length of the line. In some cases, the "smooth curve" will show a change in slope. Sometimes the change is abrupt while other times it is gradual. When looking at a temperature vs. time graph for something which cools and then starts to freeze, the graph changes from a line with a slope to a line with zero slope. One's reaction to it should be, "Oh, something important happened here," because one would expect the line to have two different slopes. Finally, when the data don't agree with the expected, it may be that you are on the road to a great discovery such as quasars, microwave background radiation from the big bang, high temperature superconductors, or many other possibilities!

Two examples of graphs are shown below. Line 1 on each of the graphs gives the best representation of the data and has points above and below for the entire length of the line. Lines 2 and 3 are less satisfactory plots of the data.

Example 1

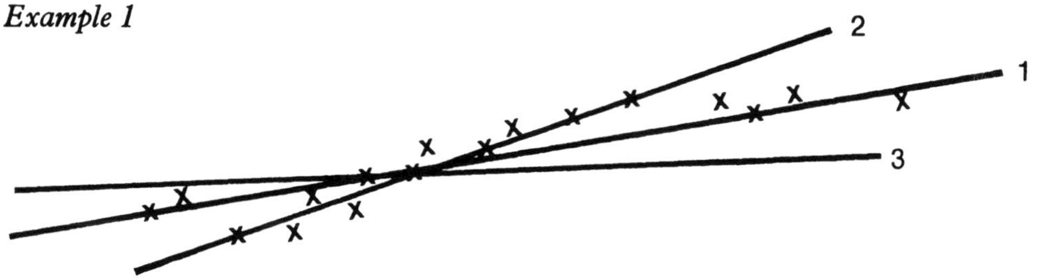

Example 2

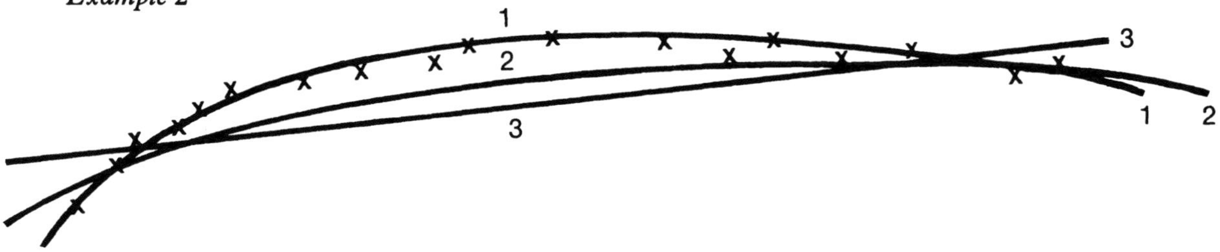

Lab #1
The Mass of Air in a Room

Name: ______________________

Date: ______________________

In this lab you will find the mass of air in your classrooom. You will measure the volume of the room and then, using the density of air, calculate the mass of the room's air. Because your measurements can never be exact, you will have some error. You will estimate this error with simple mathematics.

Materials
meter stick

Procedure

1. Use a meter stick to measure the length of the classroom three times, starting in a different place each time. Record each value. You should be able to make these measurements to the nearest .001 m (millimeter).
2. In a similar way, find the width and height of your classroom three times and record these values. Make each measurement in a different place in the room.

Analysis

1. Find the average value for the length. Record this.
2. Give an estimate of how much error there is in the length. This is the numerical error, or error in number form.
3. Find what percent the numerical error is of the length. This is called the percent error. To find what percent the numerical error is, divide the numerical error by the length and multiply by 100%.

$$\frac{\text{numerical error}}{\text{length}} \times 100\% = \text{percent error}$$

4. Repeat steps 1–3 for the width and height of the room.
5. Find the room's volume by multiplying the length times the width times the height.
6. Find the estimated error in volume. When you multiply or divide numbers, add the percent error which you have calculated.

(continued)

Lab #1

Name: ______________________

The Mass of Air in a Room *(continued)*

Date: ______________________

7. Find the mass of the air in the room using the density of air as 1.00kg/m^3 ± 5%. (The actual density can be calculated more precisely, or with less error, but it requires using the temperature, atmospheric pressure, and humidity.)
8. Give your answer with the proper number of significant digits.
9. Estimate the percent error for your mass of air. (You multiplied two numbers with error so you add the percent errors.)
10. Find the error in numerical form (the number of kilograms). Use the equation shown in step 3.

Conclusions

1. Is the total mass of air in the room a surprise to you? Explain.
2. How does the error for the volume compare with the error in the density of the air in the room? Does this mean that your errors in measuring are insignificant?
3. How would your value for mass of air differ if you took into account the objects and people in the room? Give an estimate of the percent change in your value for mass that these things would cause.

Quiz #1

The Mass of Air in a Room

Name: ______________________________

Date: ______________________________

1. The volume of a room is found by multiplying

   a. length times width.
   b. width times height.
   c. length times height.
   d. length times width times height.

2. In which of the following units is density given?

   a. m/sec
   b. m/sec^2
   c. kg/m
   d. kg/m^3

3. Pine boards have a density of 800 kg/m^3. What is the mass of 4 m^3?

   a. 200 kg
   b. 800 kg
   c. 2400 kg
   d. 3200 kg

4. What is 12% of 200?

   a. 12
   b. 24
   c. 120
   d. 2400

5. What percent of 20 is 5?

   a. 5
   b. 20
   c. 25
   d. 100

Lab #1: The Mass of Air in a Room

Teacher Notes

This lab is useful in that it introduces density, gives students an idea that air has quite a bit of mass, and gets them started with using numbers in the lab. The data gathering is rather simple so there is ample time for the students to work with their data. They can sort out problems and ask you questions. Students may also need calculators to help with the calculations.

If your classroom does not have a rectangular floor plan, you may want to just use a rectangular section of the room. If you use a room with a number of rectangular segments, instruct the students to find the mass of each segment, find the estimated error of each segment's mass, and then add the masses and the estimated numerical errors.

The rules for estimating error are: When adding or subtracting, add the numerical error. When multiplying or dividing, add the percent error. When squaring a number, multiply it by itself to double the percent error. For the reverse process, a square root, divide the percent error by two.

Possible Problems

1. Students may not like to use percents or may be fuzzy about what percent means. Many students have difficulty finding percents and finding a certain percent of a number.
2. Students may have a hard time estimating the error in their measurements. They may be tempted to say that there is no error if all of the measurements are the same. The problem is that they could get another significant digit or they should assume that they rounded off to the nearest digit. A measurement of 3.2 has two significant digits. By definition, the 2 is uncertain. The actual value might be 3.2 plus or minus .1, .2, .5, etc. At the very least, the 2 has been found by rounding. The person taking data said that it was not 3.1 or 3.3. Therefore, 3.2 has an error of $\pm$ .05. This can be used as the estimated error.
3. Some students may not like the rules for finding estimated error and may find them confusing.
4. Students may have trouble understanding that error is just an estimate of how far they think they are off. If their answer does not agree with the actual value, but is within the estimated error, then they are OK. If the actual difference is much larger than their estimated error, they should check for mistakes in the calculations or methods of collecting data.

For a more complete discussion for students on how to calculate error, see *Physics Can Be Fun: A Sourcebook of Practical Problems* (Peter Goodwin, J. Weston Walch, 1985.)

(continued)

Conclusions

1. Many students are surprised at the large mass of air. You might remind them of the force air exerts on them when they put their hand out the window of a moving car and the fact that planes fly by moving air one way or the other.
2. The measurement error may be insignificant compared to the error in density. This shows students that they can get numbers with very small percentage errors while other numbers may be more uncertain. This is important to point out to the students.
3. The objects and people might make a difference of 1–5% depending on the room.

LAB QUIZ ANSWERS

1. d
2. d
3. d
4. b
5. c

Lab #2

Acceleration of Gravity

Name: ______________________________

Date: ______________________________

In this lab you will investigate an object's acceleration under the influence of gravity. You will measure distances, calculate the average velocity for intervals of time, and then plot this average velocity on a graph of velocity vs. time. Using the graph, the acceleration due to gravity can be found.

Materials

timer	1-kg mass
strip of paper	meter stick
carbon paper disk	wooden clothespin
piece of wood to protect the floor	

Procedure

1. With the timer at the edge of a table (see Figure 2.1), take a strip of paper which is about 1.5 m long and thread it through the slot on the timer as shown in Figure 2.2.

Figure 2.1

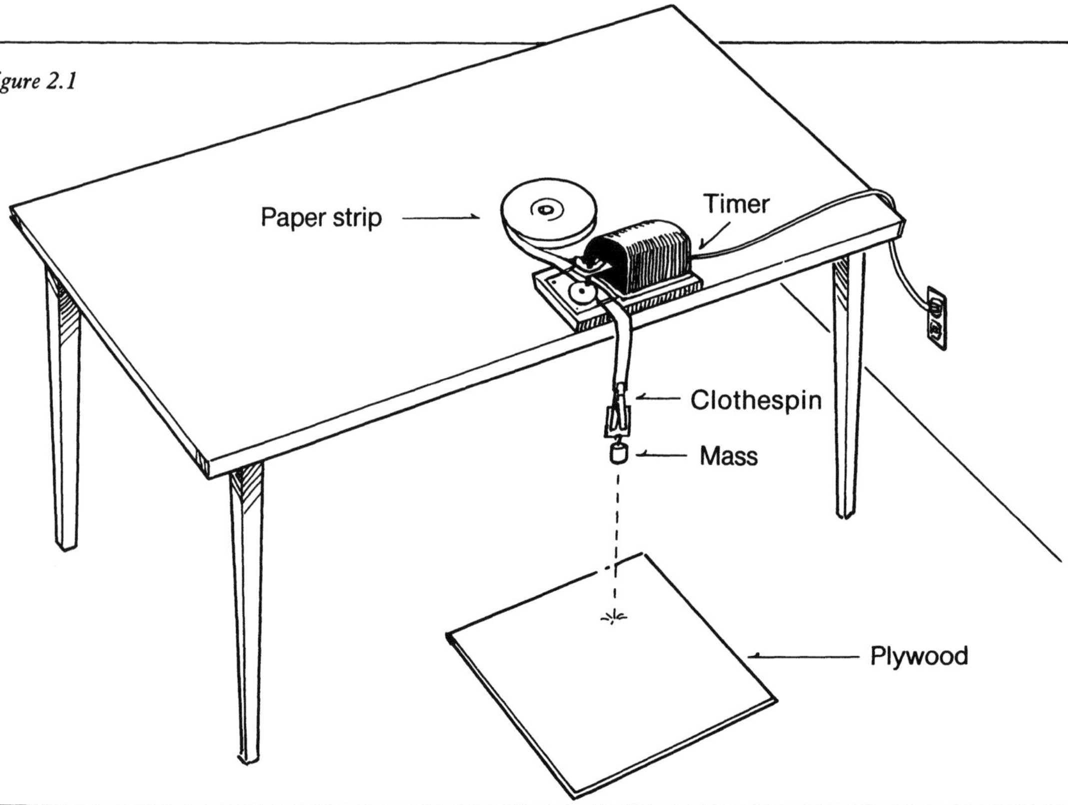

(continued)

Name: ______________________

Date: ______________________

Arrange the materials so that when the timer vibrates, the striker hits the carbon paper disk with the paper strip underneath it. This causes a dot to be marked on the strip of paper each time the timer strikes the carbon paper. If the strip is moving, the dots will be separated by a distance equal to the distance which the strip moved in one time interval.

Figure 2.2

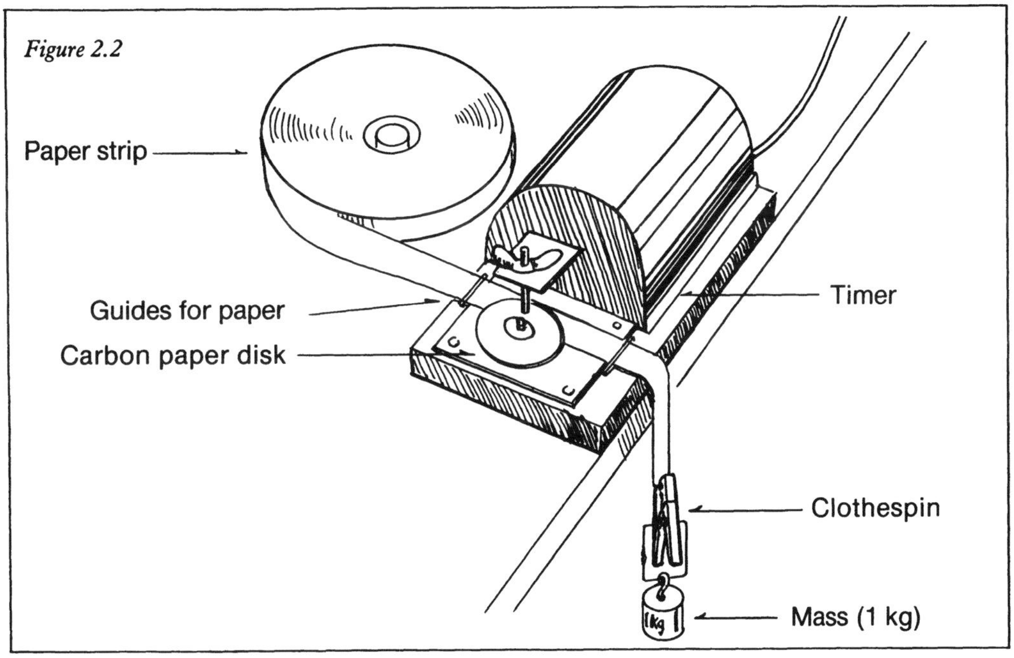

2. Place a piece of wood under the table to protect the floor when a mass falls.
3. Using a clothespin, attach a 1-kg mass to the end of the paper strip so that as the mass falls, it will pull the paper through the timer. Turn on the timer and release the mass. After the mass hits the floor, turn off the timer.
4. Mark the end of the paper strip which was attached to the mass, and remove your tape from the apparatus. As the mass dropped, the paper strip moved past the vibrating timer, and the motion of the mass was recorded.
5. Look at the dots on your strip of paper and note that the end attached to the mass has dots very close together. Then the separation becomes greater and greater. At the other end, the dots may start to be closer together because the mass hit the floor. Where increasing separation is obvious, label a dot *0*. Then number 10 or 15 more dots (*0*, *1*, *2*, *3* . . .) but make sure that the last dot is not in the area where the dots seem to get closer together. A sample is shown below.

. . . . etc.
0 1 2 3 4 5 6 7 8

(continued)

Lab #2

Acceleration of Gravity *(continued)*

Name: ______________________________

Date: ______________________________

6. With a meter stick measure how far each dot is from dot *0*. After one interval, the tape moved to position *1*. After two intervals of time, it moved to position *2*, etc. Remember to measure the distance from *0* to *1*, *0* to *2*, *0* to *3*, etc. Do *not* measure the distance between dots. Record distances for 10–15 dots in a data table similar to the one shown below. Measure these distances as precisely as possible, at least to the nearest .001 m (1 mm).

Data Table

Dot Number	Position	Change in Position	Time Interval	Average Velocity
	(m)	(m)	(sec)	(m/sec)
0	.000			
		.008	1/60	.48
1	.008			
		.010	1/60	.60
2	.018			
		.012	1/60	.72
3	.030			
		.015	1/60	.90
4	.045			

Data Table

Dot Number	Position	Change in Position	Time Interval	Average Velocity
	(m)	(m)	(sec)	(m/sec)

(continued)

Lab #2
Acceleration of Gravity *(continued)*

Name: ______________________

Date: ______________________

Cautions

1. Make sure that the paper strip moves past the timer with as little friction as possible.
2. Make sure that the wood is under the falling mass to protect the floor.

Analysis

1. Use your data to find how far the mass moved during each interval—its "change in position." Find this by subtracting the distance to *1* from the distance to *2*, etc. Record these values in your data table. (The distance traveled in the first interval is just the distance to position *1*.)
2. Use the interval of time between each vibration of the timer which is given by your teacher to calculate the average velocity for each interval. Use the distance traveled in the interval and the time that passes in each interval. Make sure that the units are distance per unit time. Record these values in your data table.
3. Draw a graph of average velocity vs. time (time on the horizontal axis) with appropriate units. Use values from your data table. After you plot your points, draw a straight line through these points. It will *not* go through all the points. Some dots will be above the line and some below.
4. Find the slope of the graph by marking two points on the line you just drew. Then find the change in velocity and divide it by the change in time (rise over run). This is the acceleration due to gravity.

Conclusions

1. Comment on how accurate your measurements were and how you could make them more precise. What quantities did you actually measure?
2. Comment on the accuracy of the experiment. Did you get the accepted value for the acceleration due to gravity? Do you think that you were lucky? If so, why?
3. The displacement measurements show that the mass is accelerating. How do they show this?
4. What does your graph show and how does it indicate accelerated motion?
5. State why the acceleration due to gravity can be found using the slope of the graph.
6. Are there any other possible sources of error? If so, what are they and how large are they?

Quiz #2
Acceleration of Gravity

Name: ______________________

Date: ______________________

1. The slope of the graph of ________________ gives you the acceleration due to gravity.

   a. displacement vs. time
   b. distance vs. time
   c. velocity vs. time
   d. velocity vs. displacement

2. The proper units for acceleration are

   a. m.
   b. m/sec.
   c. sec/m.
   d. m/sec^2.

3. After you drop the mass and it accelerates, the dots on the tape should have ________________ spaces between them.

   a. smaller and smaller
   b. equal
   c. larger and larger
   d. equal then smaller

4. If on a paper tape there were .051 m between dot *0* and dot *4* and .064 m between dot *0* and dot *5*, what was the displacement during the interval of time between *4* and *5*?

   a. .013 m
   b. .026 m
   c. .051 m
   d. .064 m

5. If two dots were .060 m apart and the time interval between dots were 1/60th of the second, what was the average velocity for the interval?

   a. .001 m/sec
   b. .060 m/sec
   c. .6 m/sec
   d. 3.6 m/sec

Lab #2: Acceleration of Gravity

Teacher Notes

Cautions

1. Make sure that the paper strip moves past the timer with as little friction as possible.
2. Make sure that the wood is under the falling mass to protect the floor.

The acceleration due to gravity is found quite precisely in this lab. As long as the tape does not get tangled as it goes through the timer, the only real error is measurement error, which can be kept low. In the lab students gather data, analyze them, and then draw a graph. The graph is used to find the acceleration due to gravity. Results generally are within 5% of the actual value of *g*.

You can set up this lab easily using an acceleration timer (Science Kit, 16025, which costs about $20), the standard paper tapes provided, and a kilogram mass with a clothespin attached to it. The timer pulses at 60 hz. You may use other timers having a fixed number of pulses per second. The actual pulse rate for other timers can be found by pulling a paper tape through the timer for a known amount of time and then counting the dots.

The clothespin attached to the 1-kg mass allows quick connection between the tape and the mass. The large mass means that frictional forces are small compared to the gravitational force on the mass. A .5-m-square piece of plywood will protect the floor when the mass hits.

Possible Problems

1. Students may not measure as directed. They may measure the distance between the dots rather than the distance from the first dot (*0*) to the indicated dot. This makes the data table fail to work as it is designed.
2. Students may not be confident about drawing graphs or how to start. They may also have difficulty figuring out the scale for the time axis which is 1/60th of a second.
3. Finding the slope of the graph and using the proper units may be difficult for some students. The slope should give the acceleration due to gravity (9.8 m/sec^2) with the proper units so it is easy to check. However, some students may be confused because they are working with variables different from those which they have used in math class (v and t as compared to x and y).

(continued)

4. Estimated error is difficult for most students and some teachers may ignore this step. However, students can give some estimate of how far they think they are off and state reasons for this. This can be done even without formal mathematical derivation of estimated error. It also helps them to understand that answers are not always exact and that if they get a value of 10.3 m/sec^2 instead of 9.8 m/sec^2, they are not wrong.
5. Students find it hard to relate the tape to what happened. At the start of the tape, many dots are together indicating the mass has not been dropped. At the end, many dots are together and are not becoming farther apart indicating the mass hit the floor. These ideas may be hard for students to visualize.

Conclusions

1. Students should be able to measure to .001 m, if not to .0001 m. As the distances get larger, percent error becomes smaller and smaller. The error in time is negligible if the timer seems to run at a constant pitch. The recommended timer works off of line frequency which makes the timing very precise.
2. Generally, the errors are less than 5 or 10% which agrees with the estimated errors in measurement.
3. The displacement increases from one interval to the next.
4. The graph shows increasing velocity as time increases, which is acceleration.
5. The slope of the v vs. t graph should give acceleration ($\Delta v/\Delta t = a$).
6. Students might suggest ways to improve the lab, such as reducing the friction as the tape moves past the timer.

LAB QUIZ ANSWERS

1. c
2. d
3. c
4. a
5. d

Lab #3
Bicycle Acceleration

Name: ______________________

Date: ______________________

In this lab you will investigate how displacement, acceleration, and time are related using a gentle hill and a bicyclist. The data which you record will help you get a better feel for the equation $s = (1/2)at^2$.

Materials	
gentle hill	bicycle
chalk	timer
meter stick	

Procedure

1. Select a paved hill which has a constant gentle slope. Use chalk to mark a finish line across the bottom of the hill.
2. Draw three more lines across the hill, one 5 m up from the finish, one 10 m up from the finish, and one 20 m up from the finish (see Figure 3). Measure these distances carefully with a meter stick.
3. Select a rider who will start from rest at the 5-m mark. Without pushing, the bicyclist will coast to the finish line. Time how long it takes to travel the 5 m. Record the time for a number of trials in a data table. Omit any trials in which the rider didn't coast in a straight line, pushed off at the start, or did anything else that would affect the data.
4. Record the time for the rider to travel 10 m and then 20 m, each time starting from rest.

Caution

Use extreme caution if you are using a hill on which there is automobile traffic.

(continued)

Name: ____________________

Date: ____________________

Figure 3

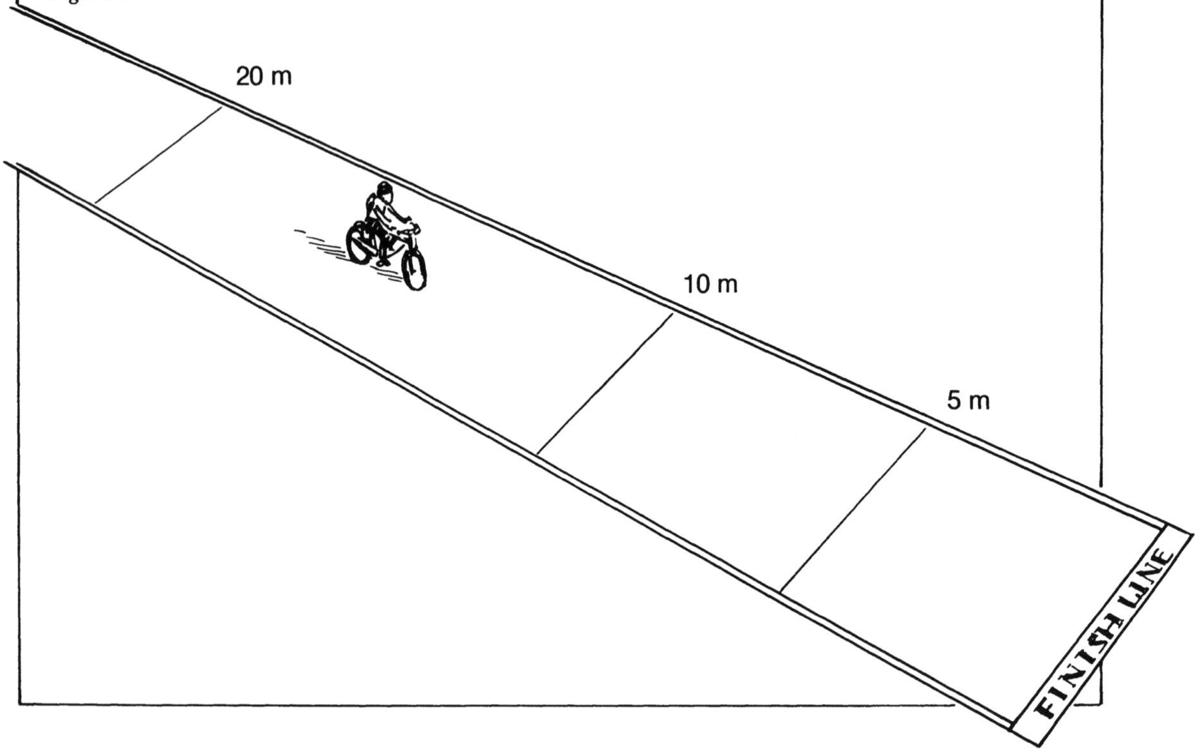

Analysis

1. First, find an average for the time it takes to go 5 m, 10 m, and 20 m starting from rest.
2. Estimate how far off, in seconds or tenths of seconds, your value for time is from the actual value.
3. Find what percent of the average value your estimated error is for each distance.
4. Using 5 m and the average time to travel the 5 m, find a value for the acceleration of the bicycle down the hill using the equation shown above.
5. Repeat step 4 for the 10-m and 20-m displacements.

Conclusions

1. A constant slope should give a constant acceleration. Does your data support or not support this? Why?

(continued)

Name: ____________________

Date: ____________________

2. Your data have some error in them. This means that you should expect some error in your results. Does your estimated error help to explain the differences in the acceleration which you found?
3. How does the average time compare for the 5-m displacement and the 20-m displacement? Would you expect this when you look at the equation? Why?
4. Does it take twice the time to go twice the displacement (10 m as opposed to 5 m) with accelerated motion? If not, why not?
5. Do your data support the equation for accelerated motion? Why?

Quiz #3
Bicycle Acceleration

Name: ______________________

Date: ______________________

1. You time how long it takes to go 5 m a number of times in order to
   a. find an average value for time.
   b. be more careful with each timing.
   c. find the displacement more precisely.
   d. do all of the above.

2. Starting from rest, if the acceleration is constant, a bicyclist will go ______________ if she accelerates for twice as long a period of time.
   a. half as far
   b. the same displacement
   c. twice as far
   d. four times as far

3. If you accelerate for 2 seconds at 3 m/sec^2, how far do you go?
   a. 2 m
   b. 3 m
   c. 6 m
   d. 12 m

4. If you go 12 m in 2 seconds, what is your acceleration if you start from rest and have constant acceleration?
   a. 2 m/sec^2
   b. 6 m/sec^2
   c. 12 m/sec^2
   d. 18 m/sec^2

5. Which of the following could be a source of error in this lab?
   a. friction
   b. measurement of displacement
   c. measurement of time
   d. all of the above

Lab #3: Bicycle Acceleration

Teacher Notes

Caution

Use extreme caution if you are using a hill on which there is automobile traffic.

In this lab students work with accelerated motion. Since this is a large-scale lab, students can see, even before they process their data, how the motion occurs and how changes in distance affect the time. The lab takes students out of the classroom which, in itself, is enjoyable. Because students can gather the data quickly, they can work on the lab with the teacher present. This helps the students to feel confident about what they are doing and makes this an ideal early lab. It is possible to be less specific with the error analysis and make the lab even easier. The numbers generally work out well so, even with only a rudimentary knowledge of error, the students can see that the equation works. The acceleration in each case should be the same.

The hill must have a constant slope and it should be paved to reduce friction. It should also be in an area where students will not have too much problem with automobile traffic. Generally, some students have stopwatches which they will eagerly use. Form as many groups as there are stopwatches as long as there are at least two students in each group. This works better than the whole class collecting the same data because it makes the numbers slightly different and prevents copying of write-ups.

The lab can be done even without a hill. A small rope and a spring scale are substituted for the force of gravity. A student pulls on a rope attached to a spring scale and keeps a constant force as the bicyclist is accelerated for the given distances.

Possible Problems

1. If the bicyclist pushes off at the start, the time will not be correct. The initial velocity must be zero.
2. Students may not use an appropriate number of significant digits and rounding errors cause errors to become large. They should have at least two signnificant digits.
3. The concept of error is hard for many students even though it just gives them an indication of when they are "in the ball park." You may want to modify the lab if you want to deal with error differently.

(continued)

Teacher Notes (continued)

Conclusions

1. The data should support the idea of constant acceleration for the different displacements, within error.
2. Yes, the error helps explain the differences in the acceleration.
3. The time should be twice as great to go 20 m as compared to 5 m. This is expected from the equation because time is squared.
4. To go twice the displacement, it takes 1.41 times longer (the square root of 2). This follows directly from the equation.
5. The data should roughly support the equation. Friction may be a factor as well as other errors associated with timing and nonconstant acceleration.

LAB QUIZ ANSWERS

1. a
2. d
3. c
4. b
5. d

Lab #4

Acceleration as Viewed with a Strobe Light

Name: ______________________

Date: ______________________

In this lab you will investigate a cart's acceleration as it moves down a ramp. You will measure displacements, calculate the average velocity for intervals of time, and then plot this average velocity on a graph of velocity vs. time. Using the graph, you will be able to determine the acceleration of the cart.

Materials

cart	film
strobe light	meter stick
camera	

A cart with a vertical marker will move down a gentle slope. A strobe light will illuminate the cart and marker as it moves and a camera will record the motion. The camera lens will be kept open during the motion so the strobe light will illuminate the vertical marker in different places as it moves down the slope. The marker will allow you to see where the cart was at different moments of time. (See Figure 4.)

The strobe light will flash at very precise intervals of time. It may be hard to find this interval of time exactly so you may be forced to use time intervals which are just "flashes" instead of fractions of seconds. This does not affect the outcome of the lab; it just means that the units of time are not seconds. Because the interval of time is going to be "flashes," it is also not important to convert displacements on the photograph to displacements in real life. Therefore, you can use "millimeters (mm) on the photo" as units of displacement. Velocity can be recorded as "mm on the photo/flash."

Procedure

1. Set up the apparatus and get ready for taking a photograph by having one person to hold the strobe light, another to hold the cart until it should be released, a person at the light switch so the room can be made dark, and another to run the camera.
2. When everyone is ready, the strobe operator turns on the strobe, the light person turns off the lights, and the cart operator releases the cart. After the lights are off, the camera operator opens the camera lens and leaves it open until the cart moves out of the picture. (While the camera lens is open, the camera must be kept motionless.) Then the lens is closed, the strobe is turned off, and the lights are turned on. The appropriate procedure is then used to develop the film.

(continued)

Lab #4

Acceleration as Viewed with a Strobe Light *(continued)*

Name: ______________________

Date: ______________________

Figure 4

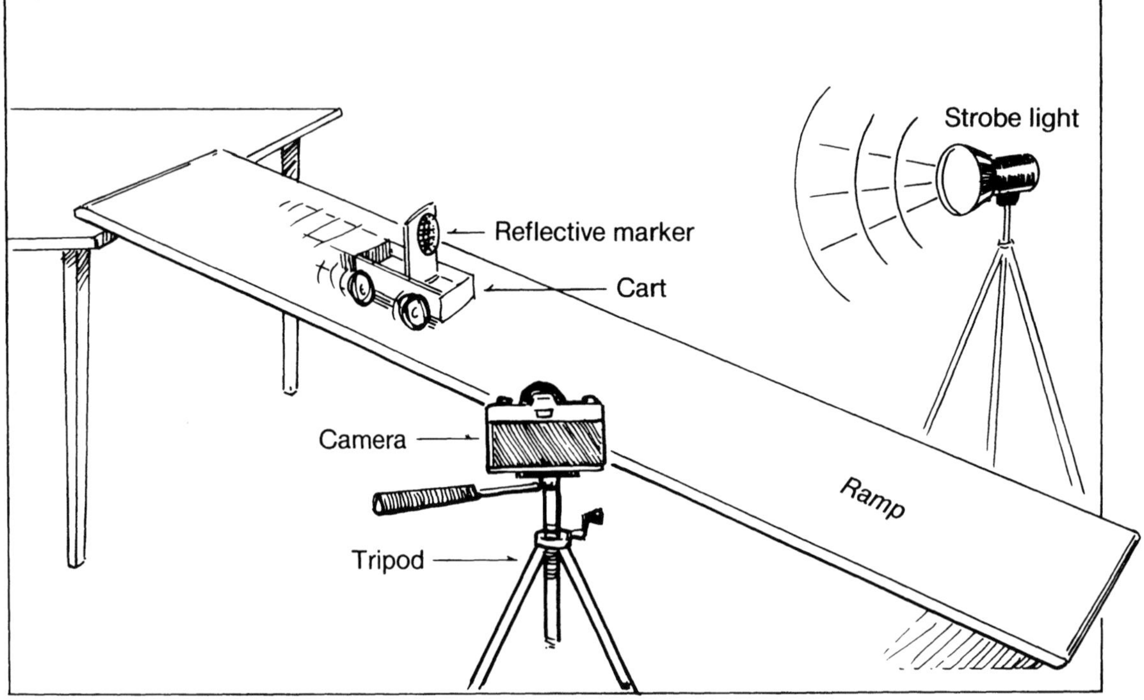

3. Look at the flashes on your photograph. The dots are very close together initially but the separation becomes larger. Record data starting where the marks are far enough apart to measure accurately. Label your starting mark *0* and then number 10 or 15 more marks (*0, 1, 2, 3* . . .). A sample is shown below.

```
. . . . . . . .   .   .    .    .     .      .       .       etc.
              0 1  2   3    4    5      6       7        8
```

4. Measure with a ruler how far each mark is from mark *0*. These measurements must be made carefully. (You must measure to the nearest .001 m but may be able to estimate to the nearest .0001 m.) Between flashes, the marker which is attached to the cart moves from one position to the next. From position *0*, after two intervals of time, it moved to position *2*, etc. Remember to measure the displacement from *0* to *1*, *0* to *2*, *0* to *3*, etc. Do *not* measure the displacement between marks. Record the displacements for 10–15 marks in a data table similar to the one shown on the next page.

(continued)

Lab #4

Acceleration as Viewed with a Strobe Light *(continued)*

Name: ______________________

Date: ______________________

Data Table

MARK NUMBER	POSITION	CHANGE IN POSITION	TIME INTERVAL	AVERAGE VELOCITY
	(m)	(m)	(sec)	(m/sec)
0	0			
		.008	1/60	.48
1	.008			
		.010	1/60	.60
2	.018			
		.013	1/60	.78
3	.031			
		.014	1/60	.84
4	.045			

Data Table

MARK NUMBER	POSITION	CHANGE IN POSITION	TIME INTERVAL	AVERAGE VELOCITY
	(m)	(m)	(sec)	(m/sec)

Cautions

1. Be careful with the photographic chemicals you use to develop the instant film. They are dangerous to handle.
2. Do NOT use strobe frequencies near 5 hz. Some people experience disturbed brain function when exposed to flashes near this frequency.

(continued)

Lab #4

Name: ______________________

Date: ______________________

Acceleration as Viewed with a Strobe Light *(continued)*

Analysis

1. Use your data to find how far the cart moved during each interval, the change in position. Find this information by subtracting the displacement to *1* from the displacement to *2*, etc. (The displacement traveled in the first interval is just the displacement to position *1*.)
2. Use the interval of time, a "flash," to calculate the average velocity for each interval in "mm in photo/flash." Record these values in your data table.
3. Using your calculated average velocities, draw a graph of velocity vs. time (time on the horizontal axis) with appropriate units. After you plot your points, draw a straight line through these points. It will *not* go through all the points; some will be above and some will be below the line.
4. Using the data which you have taken, draw a graph of displacement vs. time with time on the horizontal axis. Use the appropriate units. After you have plotted all your points, draw a line which goes through them. *Do not* "connect the dots." Draw a smooth curve. Some points will be above and some will be below the line which you draw.
5. Find the slope of graph you drew in step 4 by measuring the change in velocity and divide it by the change in time. This is the acceleration of the cart.

Conclusions

1. Comment on how accurate your measurements were and how you could make them more precise. (What quantities did you actually measure?)
2. How does your graph of displacement vs. time show that the cart accelerated?
3. How does your graph of velocity vs. time show that the cart accelerated?
4. What do you estimate your error in acceleration to be for the cart going down the slope?

Quiz #4

Acceleration as Viewed with a Strobe Light

Name: ______________________

Date: ______________________

1. In order to get a photograph of the motion of the cart down the slope, you must hold the camera lens open for
   a. just a moment.
   b. one second.
   c. five seconds.
   d. the entire time the cart is moving past the camera.

2. The graph of velocity vs. time in this lab is used to find
   a. displacement.
   b. velocity.
   c. acceleration.
   d. time.

3. Factors which might introduce error into this lab are all of the following except
   a. measurement of displacement.
   b. the strobe light.
   c. movement of the camera during the photograph.
   d. the position of the marker on the cart.

4. The velocity of a cart changes from 2 m/sec to 6 m/sec in 3 seconds. What is the acceleration in the appropriate units?
   a. .33
   b. 1.33
   c. 4
   d. 12

5. If an object goes farther the second second as compared to the first second one could assume that
   a. it is accelerating.
   b. it is decelerating.
   c. it has constant velocity.
   d. its velocity is zero.

Lab #4: Acceleration as Viewed with a Strobe Light

Teacher Notes

Cautions

1. Be careful with the photographic chemicals you use to develop the instant film. They are dangerous to handle.
2. Do NOT use strobe frequencies near 5 hz. Some people experience disturbed brain function when exposed to flashes near this frequency.

Students may not recognize at first that this lab is similar to Lab #2, which involves using a paper tape to find the acceleration due to gravity. However, as they see the similarities, they will make connections between this lab and the mass-drop lab. The mechanics of this lab are different enough, however, to require a different write-up. Since it requires some of the same graphing skills, it serves as a review of concepts while expanding lab skills.

The lab is simple if you have a strobe and a Polaroid® or other instant camera. Students may also use a spark timer which will allow them to measure the distances precisely. Or, if the ramp is quite steep, they may use a buzzer timer. Using the strobe is preferable if students are going to do Lab #12, a collision experiment on the air track. Since the collision experiment, which investigates conservation of momentum and energy, requires strobe photography, practice with the equipment will be helpful.

If students use a strobe, it will be difficult for them to use seconds as the units of time. They can photograph a moving object for a given amount of time, count the number of images from the flashes, and then use a conversion factor. However, most low-cost strobes change frequency with temperature (they warm up) so even this method does not give good results. Therefore, it is best to use time intervals of "one flash." Even though the units are non-standard, the calculations are easier. Using the conversion factors confuses the students. The nonstandard time interval means displacement is not required in standard units. Therefore, it is easiest to measure displacement in terms of millimeters on the photograph rather than trying to use conversion factors to change it into meters.

The marker on top of the cart should be reflective. It can be a small cylinder wrapped in aluminum foil taped on the top. The cart should move against a black background for the marker to stand out, and the room must be dark. It is also important to make sure that the camera lens stays open for the entire time. To make the lens stay open on some cameras, you must cover the electric eye to make it think that it is completely dark. Otherwise, you can use the setting marked "bulb." The camera must be mounted on a tripod or held firmly on a table. It helps to have the strobe aimed parallel to the track.

(continued)

Teacher Notes (continued)

You may find that it is more efficient and economical for you to take the pictures. The idea of holding the camera open for the entire event sometimes is difficult for students to grasp. (They normally take snapshots.) Therefore, you waste lots of film and time in the training process.

Because the light intensities and exposures are critical, it is best to practice the procedure before the lab. Correct exposure and a still camera are important for the outcome. The slope and/or strobe flash rate must be adjusted until there are between 10 and 20 marks on the photograph.

Possible Problems

1. Students may not measure to .001 m. Because the photograph is small, if they measure only to .01 m, their data are not meaningful. It is best if they interpolate the values between the lines on a millimeter scale. Doing this also allows them to see that measurements to .0001 m are possible with reasonable effort.
2. Students may find it hard to graph and get information from a graph. In this lab, they are directed to draw two graphs and find the slope of one of them. Some students do not have the required math skills or are not confident with this procedure. After they receive help and are on the right track, most do well.
3. They must make the measurements as directed. Many want to measure the displacement between the dots rather than from the first dot (*0*) to the indicated dot. This makes the data table fail to work as designed.
4. Some students may be confused because they are working with variables different from those which they have used in math class (v and t as compared to x and y).
5. Estimating error is difficult for most students and some teachers may wish to ignore this step. However, students should be able to estimate how far they think they are off and state reasons for this.
6. Students will try to connect the dots rather than draw a smooth curve. The smooth curve acts as an averaging process and avoids small errors in individual data points.
7. It has been suggested that prolonged flashing of a strobe at a frequency near 5 hz can cause problems in some people's brain function. Avoid frequencies near 5 hz.

Conclusions

1. The error in the smallest displacements is about 5 or 10%. For the larger displacements, the error is less than about 1 or 2%.
2. The slope of the graph of displacement vs. time keeps increasing, which indicates that the velocity keeps increasing.

(continued)

Teacher Notes (continued)

3. There is a constant slope for the graph of velocity vs. time, which indicates that the velocity increases at a relatively constant rate. Because the slope is constant, it might be assumed that acceleration is constant.
4. Time is precise as long as the strobe is working properly and the flashes are equal intervals apart. The errors in displacement might be on the order of 5 to 10%. (Unless you want to work with conversion factors, etc., you will not be able to get a "known" answer to which to compare the student's work. Because of possible variations in a noncalibrated strobe, it is probably not worth the effort.)

LAB QUIZ ANSWERS

1. d
2. c
3. d
4. b
5. a

Lab #5

Name: ______________________

A Pendulum, or How a Grandfather Clock Keeps Accurate Time

Date: ______________________

In this lab you will investigate what factors affect the period of a pendulum. The period of a pendulum is how long it takes to make one complete cycle, back and forth. Many clocks take advantage of the fact that a pendulum swings at regular intervals of time. But what factors affect how long the interval is?

Materials

string
support
paper clip
masses
protractor
meter stick
other materials found in the lab

Procedure

1. Construct a pendulum which is about 25 cm long by tying one end of the string to the support and the other end to a paper clip. Then hang a mass such as a washer on the paper clip (see Figure 5).
2. Measure the length exactly by finding the distance from the support to the *center* of the mass. Record the length in a data table similar to the one shown below.

Data Table

LENGTH	ANGLE	MASS	FACTOR 2	FACTOR 3	FACTOR 4	TIME 1	TIME 2	AVG. TIME	PERIOD

3. Pull the mass back to a small angle, measure the angle with the protractor, and record how long it takes to make 20 swings. Time to the nearest tenth of a second, if possible. Time at least twice to make sure that you got a correct time for 20 swings for this pendulum.

(continued)

Lab #5

A Pendulum, or How a Grandfather Clock Keeps Time *(continued)*

Name: ______________________

Date: ______________________

Figure 5

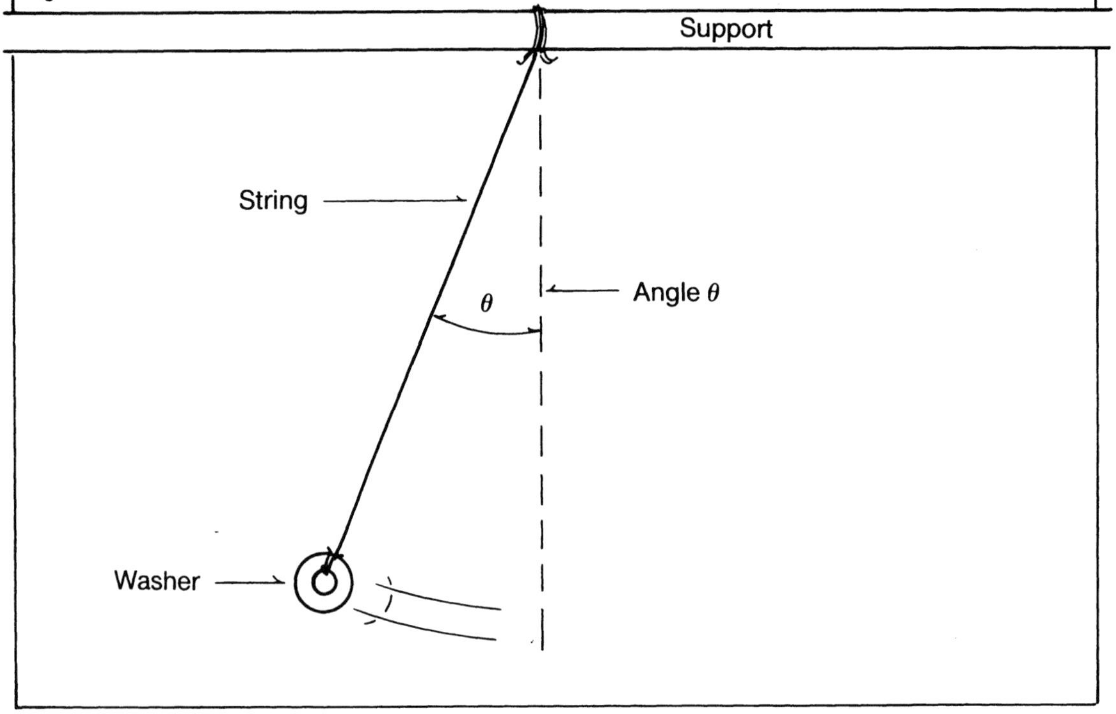

4. The pendulum you just built will be your control. You will now test what factors affect the period by changing one variable at a time. If you change the angle through which the pendulum swings, do not change any other variable. If you change more than just the angle, you cannot be sure that a change in period is caused by the change in angle or the other variable you changed. This is the function of a control. It allows you to compare your results to a known and see what factors affect the outcome of the experiment.

Investigate at least *four* factors which you think might affect the period of the pendulum. One of them should be the angle at which you allow the pendulum to swing. When you test for the angle, use a large difference such as 5, 20, and 45 degrees (not 5, 6, and 7 degrees). A large change should give you a larger change in the outcome, if one exists.

Record your data in the data table. Note that a few columns have been left blank. You can fill in these with the other factors which you are choosing to test.

(continued)

Lab #5

Name: ______________________________

A Pendulum, or How a Grandfather Clock Keeps Accurate Time *(continued)*

Date: ______________________________

If you find that one factor affects the period, investigate further. If, for instance, the angle makes a difference, find the period for at least four different angles. These will later be plotted on a graph.

> ***Caution***
>
> As you test one factor, make sure that you do not change another variable. For instance, if you attach a second mass, do not change the distance from the support to the center of the mass. If you do, you have changed two variables.

Analysis

1. First, find the period for each trial by dividing the time for 20 swings by 20.
2. Use your data for any factor which caused a change in the period to draw a graph of period vs. factor with the period on the horizontal axis. If, for instance, you found that the color of the pendulum made a difference, you would put the colors (red, orange, yellow, green, blue) on the vertical axis, and then plot the period vs. the color. You would then draw a smooth line through the points without connecting the dots. Your line should have an equal number of points above and below it.

Conclusions

1. What factors affect the period of the pendulum significantly? Give data to support this statement. How does your graph show this relationship?
2. What factors affect the period of the pendulum a little bit? Give data to support your statement.
3. What factors did not affect the period of the pendulum? Give data to support your statement.
4. Do you have questions as to whether a certain factor affects the period or not? If you do, what further tests could you make?
5. Can you think of any factors which you would like to test but could not test in the lab? What do you think would be the outcome in this "thought" experiment?
6. If you were asked to fix a grandfather clock which was gaining time, what would you do? Why?

Quiz #5

Name: ______________________

A Pendulum, or How a Grandfather Clock Keeps Accurate Time

Date: ______________________

1. Which of the following is evidence that the period of a pendulum stays constant?

   a. A grandfather clock keeps good time.
   b. A pendulum is made of simple materials.
   c. The motion of a mass is predictable.
   d. None of the above.

2. In an experiment, a control is used

   a. to keep the experiment running smoothly.
   b. for comparison.
   c. to develop a hypothesis.
   d. to gather data.

3. If a pendulum makes 10 complete cycles in 12 seconds, the period of the pendulum is

   a. .87 sec.
   b. 1 sec.
   c. 1.2 sec.
   d. 2 sec.

4. When you test for different factors affecting the period of your pendulum, you should change ______________ factor(s) at a time.

   a. one
   b. two
   c. three
   d. four

5. If you are testing to see what difference a change in angle makes and you initially use an angle of 10 degrees, which of the following would be the most helpful angle to test in your next trial?

   a. 9 degrees
   b. 11 degrees
   c. 15 degrees
   d. 45 degrees

Lab #5: A Pendulum, or How a Grandfather Clock Keeps Accurate Time

Teacher Notes

Caution

As you test one factor, make sure that you do not change another variable. For instance, if you attach a second mass, do not change the distance from the support to the center of the mass. If you do, you have changed two variables.

This lab is designed to give students experience with the factors which affect the period of a pendulum and to help them understand the use of a control. Since half of high school physics students believe that mass affects the period of a pendulum, it is useful to have them experiment to find that it has no effect. The apparatus is simple enough for them to do the experiment at home if they need to use lab time for other experiments.

Washers are good masses to use because they can be placed on the paper clips easily. The paper clips help keep the length of the pendulum constant when extra mass is added. This helps to prevent students from finding that mass has an effect when it should have none. Students may want to try using thicker string which, if not available, they can make from a number of strands of the normal string. They can test air friction by placing a piece of paper through the hole in the washer. They can test water friction in a beaker of water (although the pendulum stops quickly).

Possible Problems

1. Students may not measure the time precisely enough, which can lead to the wrong conclusions.
2. Students may change two variables at a time without knowing it. The problem is compounded because many students want mass to make a difference so if there is a change in period, they expect it and do not look for a change in length.
3. Many students "retest" things which they have proven. They find that mass does not matter when the length is 25 cm, but they then retest it with a 50-cm length. This may be a useful test if they know what they are doing, but it takes time. If they then test all angles with all combinations of mass and length, they will not have time to test a fourth factor. If they test and find that mass does not matter, encourage them to go on and test other factors unless they have extra time at the end.

(continued)

Teacher Notes (continued)

4. Graphing is hard for some students and many are not comfortable with drawing a smooth curve and want to connect the dots. The points above and below the line indicate that drawing the line is an averaging process. It can actually be a better fit of the data than any one data point.
5. The pendulum may swing in a circle instead of back and forth. This can be corrected by making sure that the supports do not move.

Conclusions

1.–3. The students should use their data to show that length affects the period greatly, angle of swing affects it slightly, frictional force (increased by adding paper or similar object to the pendulum) increases the period slightly, but mass and other factors do not affect the period. The effects of angle and friction are hard to show and students may say that they don't know if it really affects the period.

4. The changes caused by friction and angle of swing are slight so these may or may not be included as factors. More precise timing could help determine if these were factors.
5. One "thought" experiment of particular interest would be doing the lab in an orbiting spacecraft which is in free-fall. In this case, there would be no acceleration relative to the frame of the observer so the period of the pendulum would be infinite. Thought experiments are useful for they are, in part, what led Einstein to his relativity theory. It is possible to demonstrate this effect by jumping off a lab bench just as you release a pendulum. Don't actually try to time one cycle of the pendulum! The period is infinite and the landing after an infinite free-fall might exert large forces on the person who demonstrates the effect!
6. To make the grandfather clock run more slowly, you would increase the pendulum's length. The graph shows this.

LAB QUIZ ANSWERS

1. a
2. b
3. c
4. a
5. d

Lab #6
Human Horsepower

Name: ______________________

Date: ______________________

In this lab you will investigate how much power human beings can put out when going up a hill or stairs. You will time people as they gain potential energy and then calculate their power. Power equals the potential energy gained by the runner divided by the time it takes to gain that energy ($P = W/t$). You will draw two graphs and investigate the reasons for the shapes of the graphs.

Materials

a hill or stairs	meter stick
students or other runners	timer

Procedure

1. Before the lab, set up a data table similar to the one which is shown below. You can fill in some information before having people run up the hill or stairs. In the data table, include the name and mass of each runner.

Data Table

NAME	MASS	TIME	HEIGHT	POWER	POWER/KG	NOTES
	(kg)	(sec)	(m)	(watts)	(watts/kg)	

2. With a meter stick, measure the height which the runners will gain.
3. Time each runner as he or she climbs the hill or stairs (see Figure 6). Record this time. Also note any important factors which might have affected the run such as the type of footwear. This may be important when you analyze the data.

Caution

Prevent injury by advising runners not to run too fast.

(continued)

Name: ______________________

Date: ______________________

Figure 6

Analysis

1. First, calculate the power each person generates during the run. The amount of work done is given as the change in potential energy (*mgh*). Divide this work by the time for the run to find the power.
2. Find the amount of power each person generated per kilogram by dividing the person's power by his or her mass.
3. Draw two graphs, each having mass as the horizontal axis. The first graph should plot the power vs. mass. The second graph should plot the power/kg vs. mass.
4. Draw a line, either straight or curved, through your data points. Remember that this line should not connect the dots. Some points will be above the line and some will be below it.

Conclusions

1. What do these graphs show about humans and their ability to climb hills or stairs? What trends do you see?
2. Extend the line on both graphs so that you explain what people with very small or very large mass do. Explain why you drew your lines as you did.
3. Given that one horsepower equals 746 watts, how many horsepower do you put out for the climb? Do you think that you could do this for a long period of time? Why or why not?

Quiz #6
Human Horsepower

Name: ______________________

Date: ______________________

1. Power is defined as work per unit time. The units of power are
   a. sec/watt.
   b. watts/sec.
   c. joules/watt.
   d. watts.

2. The mass of each runner is required because it is related to how much ________________ is done as the runner climbs the hill.
   a. work
   b. time
   c. distance
   d. height gain

3. Because the gain in potential energy equals the work done by the runner, only the ________________ displacement(s) is/are important for calculation of the work done.
   a. horizontal
   b. vertical
   c. horizontal and vertical
   d. distance at 45 degrees from the horizontal

4. A small child runs up a hill in 20 seconds and does 4000 joules of work. The child's power output is
   a. 20 watts.
   b. 100 watts.
   c. 200 watts.
   d. 500 watts.

5. A 50-kg girl runs up a 5-meter-tall hill in 10 seconds. What is her power output? (g = 10 m/sec^2)
   a. 25 watts
   b. 250 watts
   c. 500 watts
   d. 2500 watts

Lab #6: Human Horsepower

Teacher Notes

> ***Caution***
>
> Prevent injury by advising runners not to run too fast.

This lab is useful not only because it is fun, but also because it gives students a chance to analyze data which tend to have a lot of scatter. Since people are not as predictable as a 1-kg mass falling, the results are not "perfect." Generally, however, people in better condition and with more mass have more horsepower. Students are asked to extrapolate their data to people with large and small masses. This requires them to get away from their numbers and think about what people can do. Students also must give their mass in kilograms which is a useful metric excercise. The experiment can easily be done on two flights of stairs or a 6-m-tall hill.

Possible Problems

1. Students may not understand that the data will not be "perfect." In fact, it will be perfect for the group of people who are running up the hill—it is just that people are not all predictable in their power output. The scatter is normal.
2. You need more than ten people for data points. Otherwise, trends do not really form. If you are lucky enough to teach smaller classes, you can combine the data from two classes for better graphs.
3. The graph of power/kg vs. time is actually a test of how fast someone climbs the hill. Therefore it is a fitness test.
4. Students extrapolating the curves may ignore the fact that a 10-kg child may not be able to climb the hill or stairs at all. Likewise, a person with a mass of more than 200 kg might not be able to climb either. The students, in short, may not think about reality.

Conclusions

1. The best athletes have the highest power/kg. Also, massive people need more power just to walk around so they have a larger power output.
2. Going to both large and small masses, the line will meet the horizontal axis at about 10 kg and at about 200 kg. This is because of the way people of these masses can climb hills.

(continued)

Teacher Notes (continued)

3. The answer will vary for individuals, but no person could put out one horsepower for a long period of time because the muscle groups are not large enough. Actually, one horsepower was defined using the amount of work a large draft horse could do in a day.

LAB QUIZ ANSWERS

1. d
2. a
3. b
4. c
5. b

Lab #7

Name: ______________________

Conservation of Energy: Potential into Kinetic

Date: ______________________

In this lab, you will investigate how potential energy is changed into kinetic energy. A mass falling toward the floor will pull a cart along the horizontal table. The change in potential energy ($PE = mgh$) of the falling mass shows up as kinetic energy ($KE = (1/2)mv^2$) if frictional forces are small.

Materials

cart	pulley
small masses	meter stick
balance	timer
string	

Procedure

Case 1:

1. Find the mass of the cart and mass of the object which will fall to the floor with a balance.
2. Attach a light string between the mass and cart and place the string over the pulley. The string should be long enough so that the mass can be next to the pulley and then fall to the floor without having the cart hit the pulley (see Figure 7).
3. With a meter stick, measure the distance which the mass falls as it goes from next to the pulley to the floor. Record this value.
4. Measure the time for the mass to go from the pulley to the floor. Start the clock when someone releases the cart and stop it when the mass hits the floor. Run a few trials and use the average time. This time will be used to calculate the velocity at which the cart and mass are moving.

Case 2:

5. Find the time for the mass to drop when the falling mass is doubled.

Case 3:

6. Use a cart with twice the mass of the original and the same falling mass you used in Case 2.

(continued)

Lab #7

Conservation of Energy: Potential into Kinetic *(continued)*

Name: ______________________

Date: ______________________

Figure 7

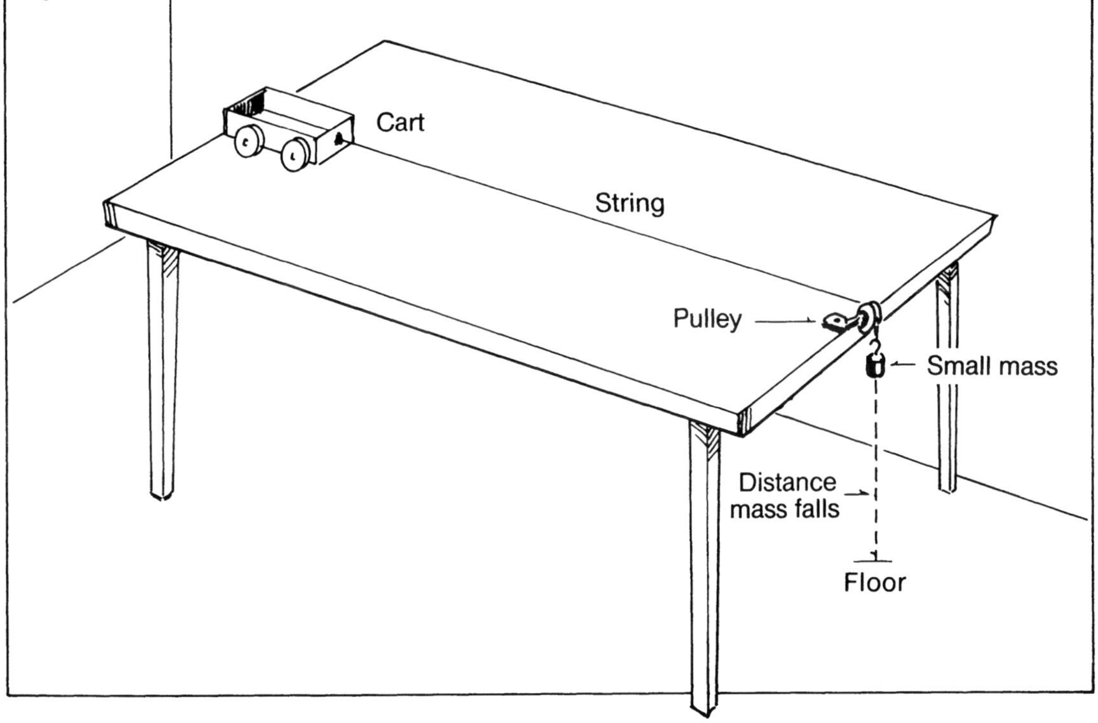

Cautions

1. Make sure that the mass drops from the same height each time.
2. Make sure, when you tie the second mass on, that the distance the mass falls is still the same as before.

Analysis

1. For the data you obtained in each case, find the average time for the mass to hit the ground.
2. Find the average velocity of the cart as it moves down the track.
3. Given that the cart starts from rest, the final velocity is twice the average velocity. (What equation shows this?) Therefore, find the final velocity that the cart and falling mass were going just before the falling mass hit the floor.

(continued)

Name: ______________________

Date: ______________________

4. Find the kinetic energy using the appropriate mass and the velocity which you just calculated.
5. Find the change in potential energy as the falling mass moved from the pulley to the floor.
6. Repeat the calculations for Case 2 and Case 3.

Conclusions

1. Was energy conserved in your experiments? Give numbers to support your statement.
2. Could errors associated with your measurement have been a factor in whether energy is conserved in this lab? Explain.
3. What other sources of error could affect this lab? Which are probably the largest? Why?
4. Given the equations for kinetic and potential energy, does the final velocity in Case 2 make sense when compared to the final velocity in Case 1?
5. Given the equations for kinetic and potential energy, does the final velocity in Case 3 make sense when compared to the final velocity in Case 1?
6. What could be done to improve the lab? Where would you make changes if you could make them?

Quiz #7

Conservation of Energy: Potential into Kinetic

Name: ______________________

Date: ______________________

1. In this lab, as a mass falls, it changes ______________ energy into ______________ energy.

   a. kinetic, potential
   b. kinetic, frictional
   c. frictional, potential
   d. potential, kinetic

2. A stopwatch is used in this lab to time what important event?

   a. The time for the cart to start moving.
   b. The time for the falling mass to hit the floor.
   c. The time for the falling mass to start moving.
   d. The time for the falling mass to stop moving after it hits the floor.

3. The velocity at which the cart and falling mass are moving just before the falling mass hits the floor is used for the calculation of

   a. potential energy.
   b. frictional work.
   c. kinetic energy.
   d. frictional energy.

4. A .1-kg mass falls .3 m to the floor. What is the change in potential energy in joules? (g = 10 m/sec^2)

   a. .03
   b. .1
   c. .3
   d. .4

5. A .5-kg cart is moving at 2 m/sec. What is its kinetic energy in joules?

   a. .1
   b. .25
   c. .5
   d. 1

Lab #7: Conservation of Energy: Potential into Kinetic

Teacher Notes

Cautions

1. Make sure that the mass drops from the same height each time.
2. Make sure, when you tie the second mass on, that the distance the mass falls is still the same as before.

In this lab, students investigate the change of potential energy into kinetic energy. The accuracy of results is dependent on friction. If large amounts of friction exist, students do not see energy conservation. If the frictional forces are small, energy is very nearly conserved. The lab also forces students to use energy equations as well as equations of motion.

The best apparatus for this experiment is an air track, which is relatively inexpensive and is useful for other labs and for many demonstrations. An air track and a small ball-bearing pulley give good results. If you have only one air track (which is quite normal), it is difficult to have each group of students do the lab during the lab period. In this case, you may wish to have students gather the data as a class and then work in pairs for the write-up.

If you do not have an air track, a dynamics cart with ball-bearing wheels works quite well. Because a dynamics cart does not keep moving as long as a cart moving on an air track, friction is a larger factor. You may want to run some experiments before the lab and find out how much energy is lost when the cart rolls a certain distance. You can then discuss the method of finding this value or just give the value to the students.

There are two methods for finding the energy lost to friction. For the first method, find a hanging mass which will keep the cart moving at a constant velocity. The gravitational force of this mass times the distance the cart moves is the energy lost. For the second method, mark the distance which the cart moves as the mass drops on the table. Then find the initial velocity needed by the cart to move this distance. The initial kinetic energy is then the amount of energy lost to friction as the cart moves.

Please note that this lab ignores the rotational kinetic energy of the pulley and/or wheels of the cart. If the radius and mass are small, this energy can be ignored and will show up as a reduced amount of energy which could be attributed to friction. Some students might notice that the spinning pulley must have some energy, and you can then help them to calculate the amount of energy which it contains.

In one system available for air tracks, air comes out of a curved path over which a paper tape moves. The air pushing on the tape makes a nearly frictionless system. It

(continued)

may also be possible to simulate the frictionless air-track surface by using small plastic spheres instead of a rolling dynamics cart (though cleaning up the beads will be time-consuming).

Possible Problems

1. You must use a small accelerating mass (but not so small that frictional forces are a large factor) or the timing error becomes large. Aim for 4 or 5 seconds as a minimum for the mass to drop for Case 1.
2. Some students may not include the falling mass in the calculation of kinetic energy. They will use only the mass of the cart. Generally, this doesn't cause too large an error because the falling mass is small compared to the cart mass.
3. Some students may not understand error and if their answers are slightly off from the accepted values, they may say that energy is not conserved and look no further.

Conclusions

1. Energy should be conserved in these experiments although frictional force may cause a loss of energy. Make sure students use numbers to support their statements. Make sure that students have used the mass of falling mass plus cart or they ignore some kinetic energy in the system. Many students will not use this mass in their calculations unless it is pointed out to them.
2. Measurement errors are a problem, especially if the time for the mass to hit the floor is short.
3. Friction and energy lost to rotational motion are the main other sources of error in this lab.
4. With two masses falling, there is twice as much kinetic energy. Kinetic energy is doubled when the velocity is larger by a factor of square root of 2.
5. With twice as much potential energy and twice as much cart mass, the velocity should be the same in Case 3 as in Case 1.
6. If friction could be reduced and the rotational energy loss reduced, the linear kinetic energy should be conserved.

LAB QUIZ ANSWERS

1. d
2. b
3. c
4. c
5. d

Lab #8
Energy and a Rubber Band

Name: ______________________

Date: ______________________

This lab will investigate how energy is changed from one form to another as a clothespin is accelerated by a rubber band. First you will find the work required to stretch the rubber band; then you will measure the height to which a clothespin rises to find how much energy it is given. You will also use a graph of force versus stretch to see if the rubber band acts as an ideal spring.

Materials

non-spring clothespin	mass hanger
meter stick	masses
rubber band	balance
support	

Caution

Even though this is a very weak "bow and arrow," if the clothespin strikes an eye, it can cause injury. Be careful! Wear safety goggles!

Procedure

1. Place the rubber band in a support so that the rubber band acts like the string of a bow. Place a meter stick next to the support so you can measure the distance you stretch the rubber band downward. (See Figure 8.)
2. Place the clothespin at the center of the rubber band. Record the position of the center of the rubber band in a data table like the one shown below. Because you have added no mass other than the clothespin, record the "mass" as zero.

Data Table

Position of Rubber Band	Change in Position	Mass	Grav. Force	Avg. Force	Work
(m)	(m)	(kg)	(N)	(N)	

(continued)

Figure 8

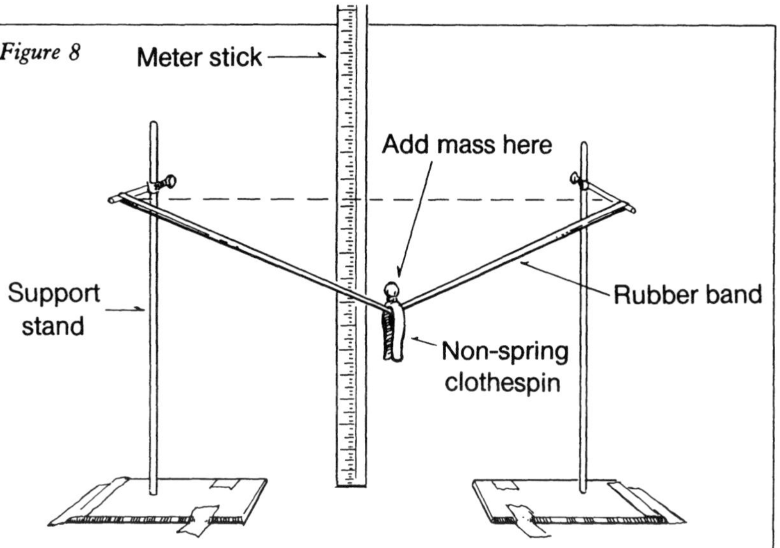

3. Now, add mass a little at a time next to the clothespin. Ideally, you should add 8–10 masses to stretch the rubber band the distance that it can be safely stretched without breaking. After adding each mass, record the total mass that has been added. Also record the position of the rubber band in your data table. Keep the clothespin on the rubber band as you add mass.
4. Remove the masses from the rubber band. Then hold the clothespin to stretch the rubber band to the position it was in with all the mass on the rubber band. Make sure that you use the meter stick to get the clothespin to the same place.
5. Release the clothespin and measure the height to which it rises. The clothespin must move straight up. The height is measured from the rubber band's position when the clothespin rests on it. Record this height.
6. Repeat step #5 a few times.
7. Find the mass of the clothespin on the balance.

Analysis

1. Find the change in position by subtracting the initial position from the final position. Record this in the data table.
2. Find the gravitational force (in Newtons) exerted by each of the massses in your data table ($f = mg$). Record these values in the data table under *force*.

(continued)

Lab #8

Name: ______________________

Energy and a Rubber Band *(continued)*

Date: ______________________

3. As the rubber band was moved, a force was exerted for a distance, but the force started at one value and increased. Therefore, average force must be used to find the work done as the rubber band is stretched. Find the average force by adding the initial force to the final force and dividing the sum by two. Find the "average" force for each time mass was added. Record these values in the data table.
4. Calculate the amount of work done as each mass was added by multiplying the average force (which you just calculated) times the change in position (in meters) which the rubber band is stretched. Record the work in the data table.
5. Find the total work done by adding the values for work you calculated in the previous step.
6. Draw a graph of force versus amount of stretch with the "amount stretched" on the horizontal axis. Use appropriate units. Draw a smooth line through these points. It may or may not be a straight line.
7. Find the amount of potential energy which the clothespin had at its highest point ($PE = mgh$).

Conclusions

1. How does the total amount of work done in stretching the rubber band compare with the potential energy of the clothespin at the highest point?
2. As you added mass, the amount of work done on the rubber band changed as mass was added. Is this change expected? Why?
3. Is the graph of force versus amount stretched a straight line? Does this mean that a rubber band acts like an ideal spring? (Remember: An ideal spring exerts twice the force when stretched twice as much.)
4. What sources of error could exist in this lab? How large are they? Which are probably the most significant?

Quiz #8
Energy and a Rubber Band

Name: ______________________

Date: ______________________

1. Work is required to stretch a rubber band because
   a. a force is exerted.
   b. a force is exerted for a distance.
   c. the rubber band moves a distance.
   d. any motion requires work to be done.

2. When the clothespin is fired, the work done to stretch the rubber band is changed into
   a. potential energy.
   b. kinetic energy.
   c. kinetic energy and then potential energy.
   d. potential energy and then kinetic energy.

3. The rubber band is stretched from .05 m to .06 m. The initial force is 1 N and the final force is 1.2 N. The amount of work done in joules is
   a. .010.
   b. .011.
   c. .012.
   d. .05.

4. In this lab, you will draw a graph of
   a. force vs. mass.
   b. force vs. stretch.
   c. stretch vs. mass.
   d. average force vs. height.

5. The setup used in this lab is a miniature version of a
   a. mass on the end of a spring.
   b. suspension bridge.
   c. bicycle wheel.
   d. bow and arrow.

Lab #8: Energy and a Rubber Band

Teacher Notes

Caution

Even though this is a very weak "bow and arrow," if the clothespin strikes an eye, it can cause injury. Be careful! Wear safety goggles!

This lab shows conservation of energy and has students find work done by a variable force. They find the work required to stretch a rubber band, then compare it to the potential energy of a clothespin at its highest point when fired by the rubber band. Generally, the results have less than 20% error.

The lab uses equipment which is only slightly dangerous and, if blunt-ended (not the spring type) clothespins are used, even the danger to an eye is reduced. It is possible to do this experiment with a bow and weighted arrow, but it is much more dangerous because the weights may come off the arrow. The lab should be done as a demonstration if a bow and arrow are used. No student should handle the bow and arrow when it is fired.

You must experiment with the clothespin, rubber band, and support so that the clothespin can be easily launched and goes to a reasonable height without the risk of breaking the rubber band. You should suggest a maximum stretch to prevent broken rubber bands. A broken rubber band means that all of the data which has been taken previously is worthless. The students must start collecting data again.

Possible Problems

1. The measurements of stretch should be done to the nearest millimeter if possible. This may be a problem because of the nature of the rubber band. Some students will find that the position changes with time. This is because work is done on the band when it is stretched. This heats it, making the rubber band tend to contract. As it cools, it relaxes.
2. The clothespin must be launched carefully so it moves straight up. There should be no sideways pull.
3. Students may have trouble doing the calculations because there are several steps. The idea of average force also may have to be explained. It may be possible to do this with a graph of force versus displacement, then talking about the area under the curve which equals work. The area of each segment of the graph is a trapezoid.

(continued)

Teacher Notes (continued)

Conclusions

1. The amount of work done on the rubber band should equal the amount of potential energy ± 20%. Height measurement and the measurement of stretch are the main sources of error.
2. As more mass is added, the force becomes larger. Thus more work is done with a given change in position.
3. The graph may or may not appear as a straight line. A straight line indicates that its behavior is like a spring.
4. Measurement of stretch and height to which the clothespin reaches are the main sources of error.

LAB QUIZ ANSWERS

1. b
2. c
3. b
4. b
5. d

Lab #9

Name: ______________________

Date: ______________________

Rope Vectors: Velocity and Displacement

In this lab you will investigate vectors and gain a feel for them. You will experience motion and displacement in two dimensions.

Materials

ropes	timer
ground limestone	bags filled with sand

Procedure

1. Mark off a square which is 30 m × 30 m on an area of grass by putting a small pile of ground limestone at each corner. Then put small piles of limestone every 2 meters along each of the four sides. The result will be similar to graph paper, but on a much larger scale. Make two sides of the square go north-south and two sides go east-west; the north-south side can be the Y-axis and the east-west side can be the X-axis. Label the origin with an "O" of ground limestone in the southwest corner. East and north are positive.
2. Have one student hold each end of each rope. Have one rope go along the Y-axis and one along the X-axis. Initially, there are two students at the origin, each holding an end of a rope. This is the "starting position."
3. Each pair holding a rope should practice walking along the sides of the square so that they walk the 30 meters in the same amount of time. They should practice until they can walk at four different speeds: .5 m/sec, 1 m/sec, 1.5 m/sec, and 2 m/sec. The timer will count out each second—"one, two, three," etc.—so the people holding each rope can figure where they should be at a particular time. Use the piles of limestone which are every 2 meters to judge the speed.
4. Once the rope operators (the people who walk holding the ropes) have practiced, return to the starting position and get a person to stand at the point where the ropes cross (see Figure 9). Have the north-south rope move 5 meters to the east. Then have the east-west rope move 5 meters north. The person where the ropes cross experiences motion east then north.
5. Return to the starting position and give the person where the ropes cross six bags of sand. Every fifth second, one bag is dropped. Both ropes should be moved at 1 m/sec. The person where the ropes cross will move north and east at the same time. You can see the path of the person where the ropes cross by noting where the bags of sand are.

(continued)

Lab #9

Name: ______________________

Rope Vectors: Velocity and Displacement *(continued)*

Date: ______________________

Figure 9

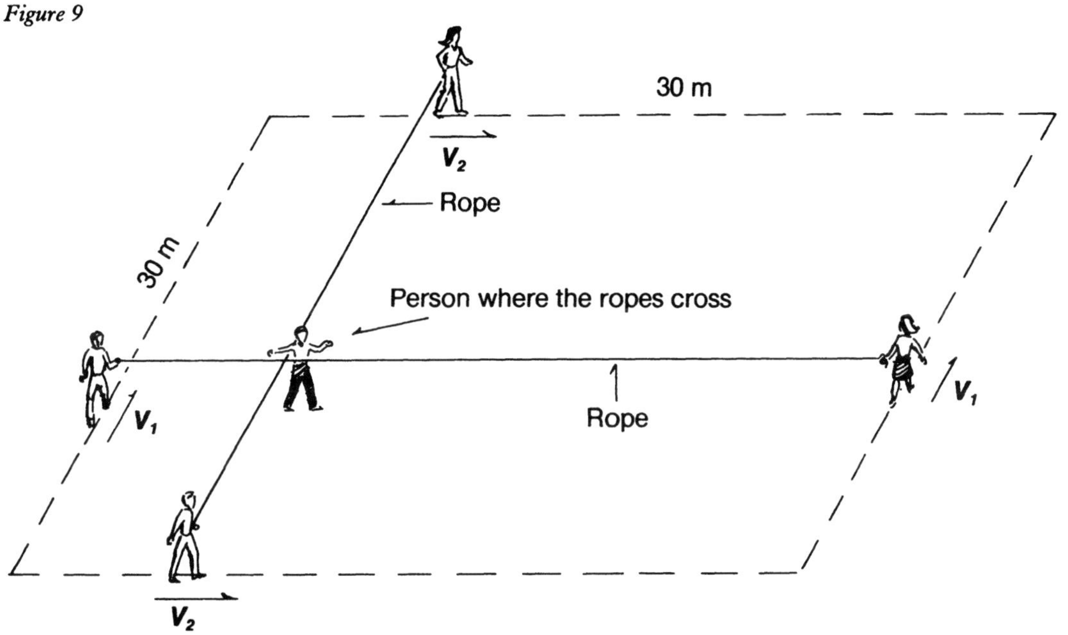

a. In what direction did the person dropping the bags move? ______________

b. How far did the person move? ______________

c. Does this agree with what the displacement should be? ______________

d. What must the velocity have been? ______________

6. Have one rope move at 1 m/sec east and the other at .5 m/sec north. Let another person be where the ropes cross and drop the sand bags.

   a. In what direction did the person dropping the bags move? ______________

   b. How far did the person move? ______________

   c. Does this agree with what the displacement should be? ______________

   d. What must the velocity have been? ______________

7. Have one rope move at 1 m/sec north and the other move at .5 m/sec east.

   a. In what direction did the person dropping the bags move? ______________

   b. How far did the person move? ______________

(continued)

Name: ______________________

Date: ______________________

c. Does this agree with what the displacement should be? ______________

d. What must the velocity have been? ______________

8. Have one rope move at 1.5 m/sec east and the other at 2 m/sec north. The resultant velocity should be 2.5 m/sec. *Note*: At 2 m/sec, you can only walk for 15 seconds before going 30 m.

   a. In what direction did the person dropping the bags move? ______________

   b. How far did the person move? ______________

   c. Does this agree with what the displacement should be? ______________

   d. What must the velocity have been? ______________

9. Move one rope at 1 m/sec east and have the other rope change its velocity. Initially, they should move at zero; by the end of 10 seconds, they should move at .5 m/sec; at the end of 20 seconds they should move at 1 m/sec, and at the end of 30 seconds should move at 1.5 m/sec. (This change in velocity may be hard to do precisely.) Again, have the person where the ropes cross drop sand bags every 5 seconds.

   a. In what direction did the person at the crossing of ropes move? ______________

   b. Does this agree with what the displacement should be? ______________

   c. What must the velocity have been? ______________

   d. Which velocity remained constant? ______________ Which velocity changed? ______________

10. Have one rope move at a constant 1 m/sec east. The other rope starts at 2 m/sec, at the end of 15 seconds has stopped, and by the end of 30 seconds is moving backward at 2 m/sec.

   a. In what direction did the person at the crossing of ropes move? ______________

   b. Does this agree with what the displacement should be? ______________

   c. What must the velocity have been? ______________

   d. Which velocity remained constant? ______________ Which velocity changed? ______________

11. In the remaining time, let the rest of the class experience the motion of where the ropes cross.

(continued)

Lab #9

Rope Vectors: Velocity and Displacement *(continued)*

Name: ______________________

Date: ______________________

Conclusions

1. Answer the questions given in the Procedure section of the lab.
2. How was the resultant velocity, when each rope moved at a constant speed, similar to the resultant velocity of two velocity vectors? Explain.
3. In step 9, one of the ropes moved faster and faster. It accelerated. How was the motion of the person where the ropes cross similar to the way that a projectile moves when fired horizontally off a cliff? Explain.
4. In step 10, how does the motion of where the ropes cross compare with the motion of a projectile fired into the air at some angle? Explain.

Quiz #9

Rope Vectors: Velocity and Displacement

Name: ____________________

Date: ____________________

1. Vectors are added together in what is called a

   a. vector sum.
   b. arithmetic sum.
   c. partial sum.
   d. complete sum.

2. The person who stands at the place where the ropes cross moves with the combination of

   a. the motion of both of the ropes.
   b. mostly the rope's motion which moves north.
   c. mostly the rope's motion which moves east.
   d. a vector subtraction of the motion.

3. If a person moves 4 m north and then 3 m east, the net displacement is ____________________ at a 37-degree angle from north.

   a. 1 m
   b. 3 m
   c. 5 m
   d. 7 m

4. When a projectile is fired horizontally from the top of a cliff, which of the following does not stay constant?

   a. acceleration
   b. horizontal velocity
   c. gravitational force
   d. vertical velocity

5. If you are traveling across a stream in a boat and are paddling at 1 m/sec as the current is moving at 1 m/sec, what is your velocity relative to shore in m/sec?

   a. 1
   b. 1.4
   c. 2
   d. 4

Lab #9: Rope Vectors: Velocity and Displacement

Teacher Notes

In this lab students visualize and feel vectors. Some students may benefit from it greatly while others will have fun and learn a little. It is important that all students, if possible, experience what it is like to be at the crossing of the ropes because that is the position where they "feel" what is happening. It is a fun lab with rough data. The write-up is brief and the lab probably can be completed in a single period if it is well organized.

The ropes can be clothesline or the equivalent. One-hundred-foot sections are about 30 m and allow you to tie small loops in each end and put markers every 2 m along the ropes. This allows the students to measure the 30-m square and 2-m grids quickly. The loops at the end allow students to hold onto the rope easily.

You may think of other possible situations which you would like to model using this form of vector addition. You may want to model the standard triangles which give easy numbers, for instance. The 3:4:5 triangle has been included, but others might be useful as well.

Be sure the students realize that the person at the crossing of the ropes must be moving at the velocity of both of the ropes, combined. This combination is a vector sum. You might point out that similar labs could be done on a river with a constant current and a constantly moving canoe. However, in this lab students are closer to the action and have much more participation in the motions. In many ways the "feel" is the important part of the lab—students can transfer this feeling to solving their vector-motion problems. Make sure they all get involved.

Possible Problems

1. Students may not be consistent in the rates at which they walk. It is easiest to check how they are doing by noting their ending positions after 30 seconds.
2. If students do not mark the grid quickly, the whole period can be wasted before any data are taken.

Procedure Answers

5. a. $30\sqrt{2}$ m
   b. NE
   c. Yes
   d. 1.4 m/sec NE

6. a. 33.5 m
   b. ENE

(continued)

Teacher Notes (continued)

c. Yes
d. 1.1 m/sec ENE

7. a. 33.5 m
   b. NNE
   c. Yes
   d. 1.1 m/sec NNE
8. a. 37.5 m
   b. at 37 degrees east of north
   c. Yes
   d. 2.5 m/sec at 37 degrees east of north
9. a. It is changing; it is a parabola.
   b. Yes
   c. It is changing.
   d. The eastward velocity stays constant; the northerly changes.
10. a. It is changing; it is a parabola.
    b. Yes
    c. It is changing.
    d. The eastward velocity stays constant; the northerly changes.

Conclusions

2. The motion of the person where the ropes cross is the same as adding velocity vectors. The person at the rope crossing must move with two components of velocity.
3. One velocity stayed constant, which is the horizontal in projectile motion, while another changed at a constant rate.
4. One velocity stayed constant, which is the horizontal in projectile motion, while another changed at a constant rate.

LAB QUIZ ANSWERS

1. a
2. a
3. c
4. d
5. b

Lab #10
Firing a Projectile Horizontally

Name: ______________________

Date: ______________________

In this lab students will investigate the flight of a projectile when fired off a table horizontally. The horizontal distance which it flies before it hits the floor will be measured and compared to the theoretical distance it should fly.

Materials

ramp	paper
clamps	carbon paper
ball	string
meter stick	washer
tape	pencil

Procedure

1. Attach the ramp to a table with clamps (see Figure 10.1). Have the lower end of the ramp slightly off the edge of the table. Make sure that a ball can roll down the center of the ramp and not hit anything. Also, make sure that the ramp is horizontal at the table edge.
2. With a meter stick, measure the height of the end of the ramp above the floor. Record this.
3. Place the ball at the highest point on the ramp where it can still roll down when released. Measure and record the height of the ball above the end of the ramp.
4. Release the ball and note approximately where it lands. Repeat this to make sure that it did not happen by chance.
5. Tape a piece of white paper to the floor with its center approximately where the ball landed. Then place a piece of carbon paper over the white paper.
6. Roll the ball down the ramp from the top of the ramp. When it hits the floor, it should hit the carbon paper and leave a dot on your white paper. Roll the ball at least five times down the ramp.
7. Take a piece of string which is long enough to reach from the end of the ramp to the floor and tie a washer to it. Hold the string at the end of the ramp so that the washer is just off the floor. Gravity will make the washer hang vertically to show the place which is under the ramp. Using a pencil, mark this point with an *X*.

(continued)

Lab #10
Firing a Projectile Horizontally *(continued)*

Name: ____________________

Date: ____________________

8. From the point you just marked with an X, measure the distance to each of the black marks made by the ball on your white paper.

Figure 10.1

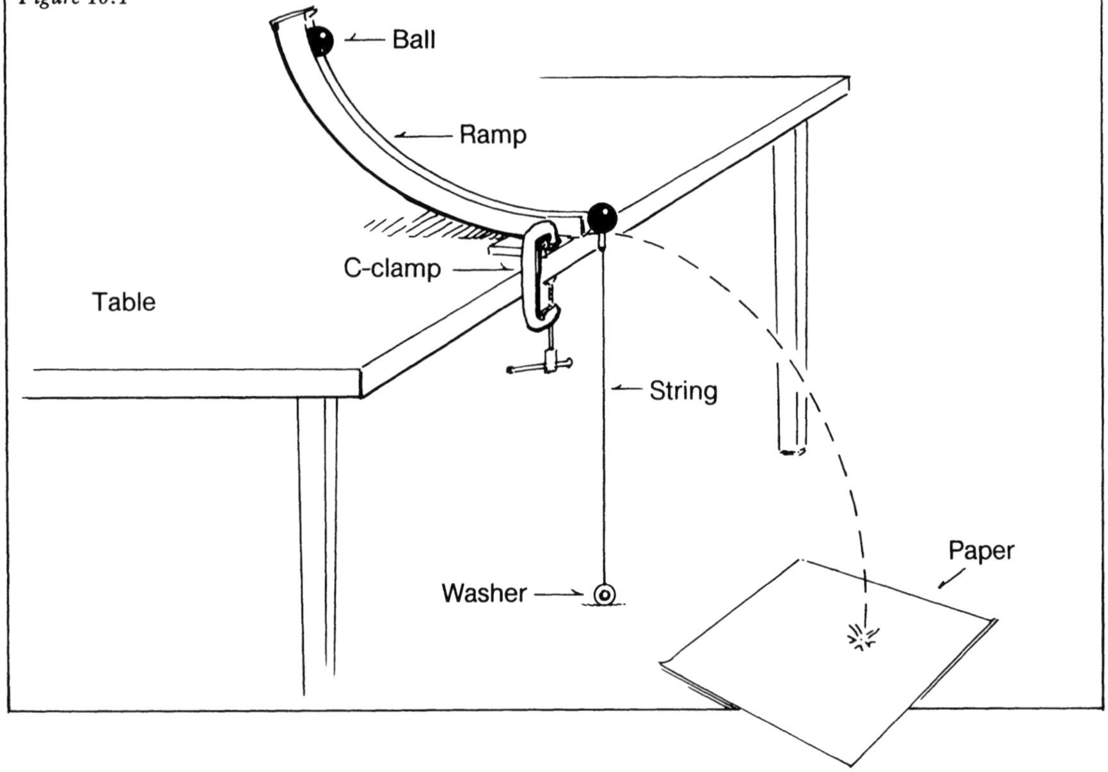

Cautions

1. Make sure that the ramp is pointed horizontally at the bottom. Otherwise, the ball will have some vertical velocity.
2. Make sure that you do not remove your piece of paper taped to the floor until you have all the proper measurements. Once it is moved, you cannot use it to take data.

(continued)

Lab #10
Firing a Projectile Horizontally
(continued)

Name: ______________________

Date: ______________________

Analysis

1. Find the average distance that each black mark was from under the ramp. Give some estimate of the error in this value.
2. Using the height of the ramp above the floor, find the time for the ball to fall from ramp to the floor. Assume that the vertical velocity is initially zero.
3. The ball moved horizontally the distance you found in step 1 in the time you found in step 2. Use these values to calculate the ball's velocity when it left the ramp.
4. The potential energy which the ball has is changed into linear kinetic energy *plus* rotational kinetic energy. It can be shown that 2/7 of the energy is in the form of rotational kinetic energy and 5/7 is in the form of linear kinetic energy (kinetic energy associated with a mass moving in a straight line). Therefore,

   $$(5/7)mgh = (1/2)mv^2$$

   Use this equation to find the velocity which the ball was going when it left the ramp.

Conclusions

1. How does the velocity found by measuring the distance the ball flew and the time it was in the air compare with the velocity calculated by using energy? Can the differences be explained by measurement errors in the lab? Can any other sources of error help cause the differences?
2. What effect on the lab would there be if the ramp were not horizontal at the lower end? How would you have to change your calculations?
3. If you are familiar with the equations for rotational motion, show why only 5/7 of potential energy is changed into linear kinetic energy.
4. Does it make sense that a rotating object takes some of the potential energy just to spin? Explain your answer.

Quiz #10
Firing a Projectile Horizontally

Name: ______________________________

Date: ______________________________

1. When a projectile is fired horizontally across a level surface, it will stay in the air as long as

   a. an object thrown up at the same velocity.
   b. an object thrown down at the same velocity.
   c. an object dropped from rest from the same height.
   d. an object dropped from a height a little bit lower than the projectile was fired.

2. The ball in this lab has kinetic energy at the lower end of the ramp because it

   a. had potential energy at the top of the ramp.
   b. is given a push.
   c. is pushed for a distance by a spring which exerts a force on it.
   d. has changed kinetic energy into potential energy.

3. If an object flies horizontally off the end of a ramp and goes 2 meters horizontally in .5 sec, what is its horizontal velocity in m/sec?

   a. .5
   b. 1
   c. 2
   d. 4

4. A ball falls from rest a distance of 5 meters. About how many seconds does it take to fall this distance?

   a. .5
   b. 1
   c. 1.5
   d. 2

5. The carbon paper and white piece of paper are used to

   a. make a second copy of your data.
   b. measure the height of the ramp.
   c. find where the ball lands on the floor.
   d. find the height of the ramp off the floor.

Lab #10: Firing a Projectile Horizontally

Teacher Notes

Cautions

1. Make sure that the ramp is pointed horizontally at the bottom. Otherwise, the ball will have some vertical velocity.
2. Make sure that you do not remove your piece of paper taped to the floor until you have all the proper measurements. Once it is moved, you cannot use it to take data.

This is a simple version of a projectile motion problem because the initial velocity is horizontal. Therefore, the time in the air can be found by finding how long it takes anything to fall from the table to the floor ($h = (1/2)gt^2$). The only problem with this lab is that not all of the potential energy of the ball is changed into translational kinetic energy because the ball is rolling. If students have worked with rotational motion equations, they should derive the equation used to find the velocity of the ball.

This apparatus is available from some supply houses in a form which also allows you to do experiments with collisions in two dimensions. The "collision in two dimension" apparatus costs $15–$20.

You can easily build the apparatus but you must make sure that it causes the ball to be moving horizontally when it leaves the ramp. Otherwise, the assumption that the vertical velocity is initially zero will not work. Construct the apparatus by screwing or nailing together two pieces of wood at right angles. Use two small nails to hold a plastic ruler in place. Position the nails so that when the ruler is put in place, its ends are flush with the board and the ball can roll down the center of the ruler and go between the nails at the end. The nails at the bottom end should be very close to the end of the support so that the ball flies off the ruler and does not hit the board. The board should be wide enough so that it can be clamped to the table top. See Figure 10.2.

Possible Problems

1. If the students move their paper before they have taken their measurements, they lose their data.
2. Students may be confused by the fact that some potential energy is changed into rotational motion.
3. If the ramps do not send the ball horizontally off the table, the data will be poor.

(continued)

Figure 10.2

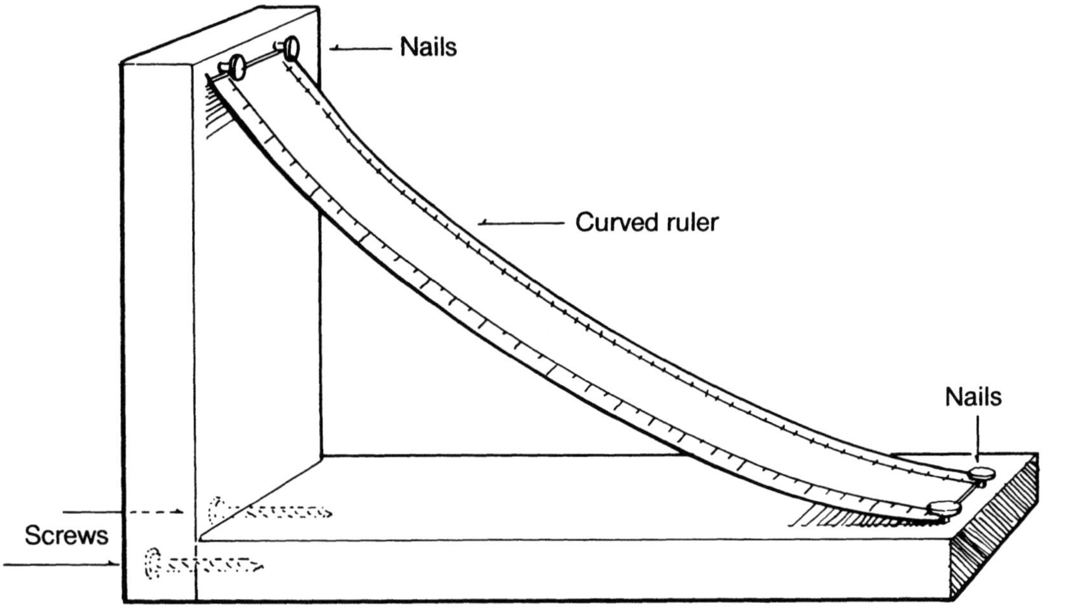

Conclusions

1. The velocities should be approximately equal although some energy may be lost as the ball slips on the ramp before it starts to roll. If you are using the metal track instead of the ruler, the ball is not rolling at a distance of 1 radius from the axis. The ball curves below the edges of the track. This should be a small enough error so that it should not be a problem.
2. If the ramp doesn't send the ball horizontally, then the vertical velocity cannot be assumed to be zero.
3. Here is a derivation from the equations for rotational motion:

$$mgh = (1/2)mv^2 + (1/2)I\omega^2$$

But I and ω are related to m and v so by substitution:

$$mgh = (1/2)mv^2 + (1/2)(2/5)mr^2\left(\frac{v^2}{r^2}\right)$$

This reduces to:

$$mgh = (1/2)mv^2 + (1/2)(2/5)mv^2 = (7/10)mv^2$$

(continued)

Teacher Notes (continued)

Therefore, 5/7 of the energy is due to linear motion, while 2/7 is due to rotational motion.

4. Any time a mass is moving, it has kinetic energy. Mass rotating on an axis is no different. We also know that a spinning object, like a bicycle wheel, has energy and can exert force for a distance. You experience this any time you put your fingers against the spokes.

LAB QUIZ ANSWERS

1. c
2. a
3. d
4. b
5. c

Lab #11
Dart Guns

Name: ____________________

Date: ____________________

In this lab you will investigate projectile motion with the help of a toy dart gun. You will find the velocity of the dart leaving the gun and then fire the dart at a known angle. You will then use the equations for projectile motion to see if they predict the actual displacement the dart flies before it hits the ground.

Materials

dart gun
protractor
meter stick

Procedure

1. Fire your dart gun vertically three or more times to get values for how far the dart rises before coming back down. Record these values in a data table. Make this measurement as carefully as possible because it will be used to find the velocity at which the dart leaves the gun.
2. Choose an angle between 20 and 70 degrees from the horizontal. Using a protractor, fire your dart three or more times from a level surface (See Figure 11). Fire at the same angle each time and measure the angle precisely. Measure the horizontal displacement from where the dart is fired to where it lands each time. Record these values.

Caution

Although you are using toy dart guns, it is easy to injure an eye with them. Do not fire darts at people. Wear safety goggles!

Analysis

1. Find the average value for the height to which the dart rose. Then find the average displacement that the dart flew horizontally. Give a value for the estimated error for each.

(continued)

Name: ______________________

Date: ______________________

Figure 11

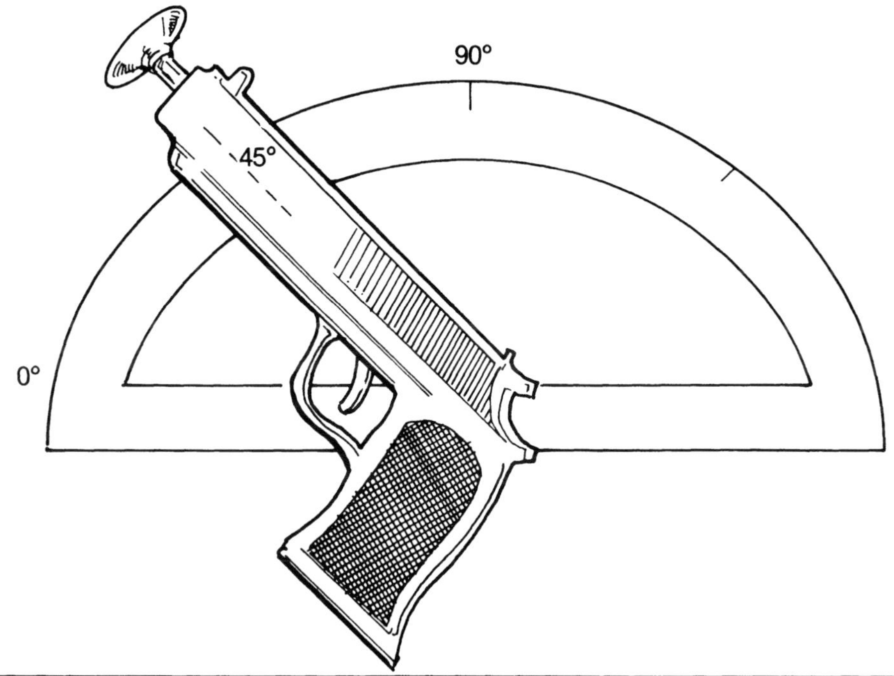

2. Calculate the velocity of the dart leaving the gun by using the height to which the dart rose and setting the initial kinetic energy equal to the potential energy at the highest point ($KE = PE$).
3. Use the velocity of the dart leaving the gun to find how far the dart should fly when fired at the angle you selected. (Find the horizontal and vertical components, find the time in the air, and then find how far it flies during that time.) Remember that the dart flies horizontally at a constant velocity for the entire time it is in the air.

Conclusions

1. Compare the actual displacement to the calculated displacement. Do the values agree, given the error which you estimated in your measurements? Support your answer with numbers.
2. What other sources of error could exist in this lab to make the dart not fly as the equations predicted? Why are they a factor?
3. Was this procedure successful in showing that the equations for projectile motion work at least in one case?

Quiz #11
Dart Guns

Name: ______________________________

Date: ______________________________

1. This lab will examine how the horizontal displacement of a dart, found by equations, compares with the
   a. theoretical height to which the dart rises.
   b. actual horizontal displacement of the dart.
   c. actual height to which the dart rises.
   d. actual distance the dart traveled when fired straight up.

2. The height measurement in this lab is used in the equation for ____________________ to find the velocity of the dart leaving the gun.
   a. potential energy
   b. kinetic energy
   c. displacement
   d. average velocity

3. After the dart is fired, the ____________________ velocity stays constant.
   a. average
   b. final
   c. vertical
   d. horizontal

4. After the dart is fired, the ____________________ velocity changes at a constant rate.
   a. average
   b. final
   c. vertical
   d. horizontal

5. The dart will have a constant horizontal velocity
   a. only for the first half of the flight.
   b. only for the second half of the flight.
   c. from when it is fired until it strikes the ground.
   d. only when it is going up.

Lab #11: Dart Guns

Teacher Notes

> ***Caution***
>
> Although you are using toy dart guns, it is easy to injure an eye with them. Do not fire darts at other people. Wear safety goggles!

This lab is a classic and generally verifies the equations for projectile motion with less than about 10% error. The apparatus is inexpensive and can be found in toy stores. The best dart guns have rigid shafts but, because these are dangerous for children, manufacturers are now producing guns with soft shafts which propel darts with air. These guns work quite well but you must make sure that the darts stay in good condition so that the air will push them out of the gun reliably.

If the dart is unweighted, it may rise too far and hit the ceiling, which makes the measurement of height impossible. Ideally, the dart should rise between one and two meters when fired vertically. Weights can be added by taping washers to the front of the dart.

This lab can be simplified by having the students fire the gun horizontally from a table top. This reduces the need for using sines and cosines or similar triangles. Only slight changes in the procedure are required for this change in the lab.

Possible Problems

1. Students must carefully measure the height to which the dart rises. Ideally, you want to measure the rise of the center of mass. It is quite easy to make this measurement against a wall.
2. Using the equations is difficult for many students. It is sometimes surprising how students can do problems on the homework just like this lab but not be able to do the lab calculations.
3. If the results do not come out well, the students are liable to think that the equations do not work. Generally, it is a problem with their calculations, but there may be some problems with the gun. Sometimes the friction against the dart as it is fired at an angle is different when the gun is fired vertically. You may want to check this with the dart guns which you use.
4. Errors in angle can be lessened if students fire the dart at 45 degrees. This is because the range changes little when the dart is fired at plus or minus 5 degrees from this angle.

(continued)

Teacher Notes (continued)

5. Students may ask for the mass of the dart before they realize that mass cancels from the conservation of energy equation.

Conclusions

1. The theoretical displacement found using equations should equal the actual displacement with about 10% error. If the answers are way off, it may mean that the student has made calculation errors. There may be a problem with friction if the actual values seem to average slightly smaller than the theoretical value.
2. Students should mention friction as the dart leaves the gun as well as error in the angle fired. Very capable students might do a study by running numbers through the equations to see how angle would affect their answers. It might make a good computer project.
3. The lab should be successful.

LAB QUIZ ANSWERS

1. b
2. a
3. d
4. c
5. c

Lab #12

Collisions: Conserving Energy and Momentum

Name: ______________________________

Date: ______________________________

In this lab you will investigate if energy and momentum are conserved in an elastic collision. You will use an air track with two carts, one more massive than the other. The smaller cart is initially stationary and is hit by the larger cart. You will use a camera with a strobe light to record the motion of the carts before and after the collision. Using data gathered from the photograph, you can determine velocities and will see if kinetic energy ($KE = (1/2)mv^2$) and momentum ($p = mv$) are conserved.

Materials	
air track carts	strobe light
shiny markers	instant camera
balance	meter stick

Procedure

1. Attach a marker vertically to each cart with one on the top and one on the side (see Figure 12). The markers will allow you to see the motion of the carts and, because one is higher than the other, you can tell which marker is associated with each cart. Make sure that the track is level.

Figure 12

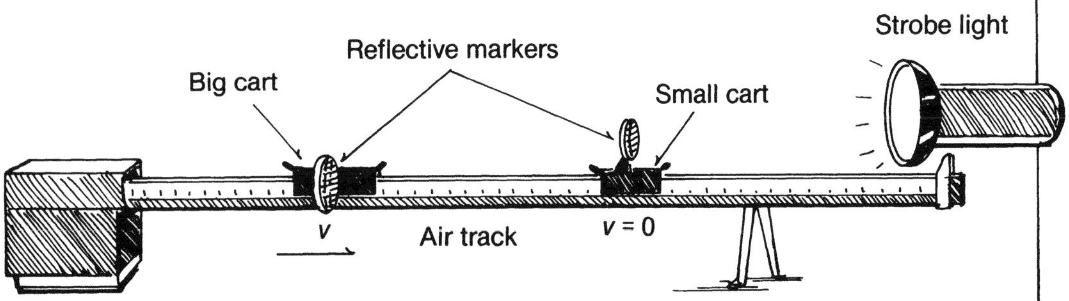

2. Find the mass of the two carts with a balance and record the values.

(continued)

Lab #12

Name: ____________________

Date: ____________________

Collisions: Conserving Energy and Momentum *(continued)*

3. Position the strobe light so it will illuminate the markers as they move along the track and position the camera so that it can record the collision. Keep the camera lens open for a few seconds before and after the collision.

The strobe light flashes at very precise intervals of time. It may be hard to find this interval of time exactly so you may be forced to use time intervals which are just "flashes" instead of fractions of seconds. This does not affect the outcome of the lab; it just means that the units which you use are not the standard units of m/sec. Because the interval of time is "flashes," it is not important to convert displacements on the photograph to displacements in real life. Therefore, you can use "millimeters (mm) on the photo" as units of displacement. Velocity will be recorded as "mm on the photo/flash."

4. To do the lab, you need one person to turn on the air track, one to hold the strobe light, another to hold the small cart stationary, a person to turn off the lights, a person to push the big cart, and another to operate the camera. You may need the help of another lab group to do all these tasks. When everyone is ready, the air track and strobe are turned on, the light person turns off the lights, the small cart is released so it remains stationary, and the camera operator opens the camera lens as the large cart is pushed. The camera lens is left open until a few seconds after the collision occurs. After the picture is taken, turn everything off (except for perhaps the strobe light if other groups are going to use it) and use the appropriate procedure to develop the film.
5. Look at the photograph. The marker for the small cart goes a constant distance between flashes. The marker for the large cart seems to go at a constant velocity and then the velocity changes to a slower one. This change represents the change in velocity from before collision to after the collision.

To find the velocities of the carts before and after the collision, look at the photograph and find where the markers seem to be a constant distance apart. For each velocity, use as many intervals as possible. Mark on the photograph which markers you are going to use. Have your teacher check these marks to make sure that you have done it properly. A sample is shown below.

The "↑" shows the location of each marker.

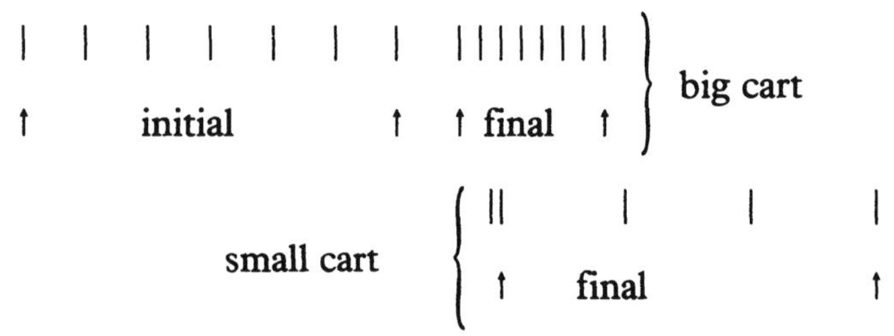

(continued)

Name: ______________________________

Date: ______________________________

6. For each velocity, measure the distance between your marks (↑) to the nearest millimeter. Then count the number of intervals. Record these values.

Cautions

1. Be careful with the photographic chemicals which are used to develop the instant film. They are dangerous and should not get on your skin or clothes.
2. Be careful not to move the camera while the photo is being taken.
3. Make sure the small cart is initially stationary.
4. Do NOT use strobe frequencies near 5 hz. Some people experience disturbed brain function when exposed to flashes near this frequency.

Analysis

1. Find the initial and final velocity of the big cart and the final velocity of the small cart by using the distances you measured and the number of intervals you counted. Make sure these velocities are labeled properly.
2. Using the mass of each cart and the velocities you just found, find the momentum of each cart before and after the collision.
3. Using the mass of each cart and the velocities you just found, find the kinetic energy of each cart before and after the collision.
4. Give an estimate of error based on your measurement errors for each of these values.
5. Find the total momentum and total kinetic energy both before and after the collision. This includes the momentum or kinetic energy from both carts.

Conclusions

1. Is momentum conserved in this experiment, given your estimate of error? Support your answer with numbers.
2. Is kinetic energy conserved in this experiment, given your estimate of error? Support your answer with numbers.
3. Is this an elastic collision between the two carts?
4. What other sources of error could occur in this lab?
5. If energy was gained or lost, where could it come from or go to? Explain.

Quiz #12

Name: ______________________

Collisions: Conserving Energy and Momentum

Date: ______________________

1. What is conserved in an elastic collision?
   a. momentum only
   b. kinetic energy only
   c. momentum and kinetic energy
   d. neither momentum nor kinetic energy

2. What statement best describes this experiment?
   a. Two moving carts collide on the air track and the event is recorded with a strobe photograph.
   b. A moving cart hits a stationary cart and the event is recorded with a strobe photograph.
   c. A moving cart collides with the end of the air track and the event is recorded with a strobe photograph.
   d. Elastic bands are used to cause an air track cart to move and the event is recorded with a strobe photograph.

3. The position of a cart is shown by *x*'s below. The distance from the first *x* to the last is 12 cm. One interval of time passes between each *x*. What is the velocity in cm/interval of time?

   x   x   x   x

   a. 1
   b. 2
   c. 3
   d. 4

4. A 2-kg cart is moving at .4 m/sec. How much momentum does it have in kg-m/sec?
   a. .1
   b. .8
   c. 2
   d. 2.4

5. A 2-kg cart is moving at .4 m/sec. How much kinetic energy does it have in joules?
   a. .16
   b. .4
   c. 1
   d. 2.4

Lab #12: Collisions: Conserving Energy and Momentum

Teacher Notes

Cautions

1. Be careful with the photographic chemicals which are used to develop the instant film. They are dangerous and should not get on your skin or clothes.
2. Be careful not to move the camera while the photo is being taken.
3. Make sure the small cart is initially stationary.
4. Do NOT use strobe frequencies near 5 hz. Some people experience disturbed brain function when exposed to flashes near this frequency.

This lab shows conservation of energy and momentum with less than 10% error in almost all cases as long as the collision is gentle and the carts do not jostle around enough to hit the air track. The lab is simple to do if you have an air track, strobe, and instant camera. Without these tools, the lab is difficult to run with satisfactory results.

The cart which is initially moving should have about twice the mass of the other cart. This means that the big cart continues moving in its original direction after the collision. The markers on the carts should be reflective. They can be small cylinders wrapped in aluminum foil taped to the carts. One must be at a lower level than the other so that in the photograph students can tell them apart. The photograph must be made against a black background or the markers will not stand out. The room must be dark. It is also important to make sure that the camera lens stays open for the entire time. To make the lens stay open on some cameras, you must cover the electric eye to make it think that it is completely dark. Otherwise, you can use the setting marked "bulb."

You may find that it is more efficient and economical for you to take the pictures. The idea of holding the camera open for the entire event sometimes is difficult for students to grasp. (They normally take snapshots.) Therefore, you waste lots of film and time in the training process.

Because the light intensities and exposures are critical, it is best to practice the procedure before the lab. Correct exposure and a still camera are important for the outcome. You want to make sure that there are at least three or four intervals for the calculation of each velocity. Fewer than this introduces measurement error. This requires adjustment of the strobe flash rate until there are this many marks on the photograph. It may also help to aim the strobe almost parallel to the track.

(continued)

One problem which occurs in this lab is that unless the strobe is accurately calibrated, it is difficult to use seconds as the units of time. It is possible to photograph a moving object for a given amount of time, count the number of images from the flashes, and then use a conversion factor. However, most low-cost strobes change frequency with temperature (they warm up), so even this method does not give good results. Therefore, it is best to just use time intervals of "one flash." This means that the units are nonstandard but it is easier to do the calculations.

Given that the time interval is not standard, it then becomes obvious that the measurement of displacement is not required in standard units. It is therefore easiest to measure displacement in terms of millimeters on the photograph rather than trying to use conversion factors to change it into meters. The units of energy and momentum will definitely not be standard, but energy and momentum should be conserved with any units which you use.

Possible Problems

1. Students may not measure to parts of millimeters. Because the photograph is small, if they measure only to millimeters, their data are not as good as if they interpolate the values between the lines on the millimeter scale. Doing this also allows them to see that such measurements are possible with a reasonable effort.
2. Students must push the cart gently so that the collision does not jostle the carts causing them to hit the track.
3. Some students may find it hard to understand the photograph. The small cart marker appears to start moving in a completely different place from where the large cart marker changed its velocity. These students need to be shown that the markers do not collide; it is only the ends of the cart which collide.
4. Estimated error is difficult for most students and some teachers may ignore this step. However, students should be able to give some estimate of how far they think they are off and state reasons for this.
5. Students will not keep track of which velocity is which. This means that they might use a final velocity to calculate initial momentum.
6. It has been suggested that prolonged flashing of a strobe at a frequency near 5 hz can cause problems in some people's brain function. Avoid frequencies near this.

Conclusions

1. Momentum should be conserved with 10% error or less.
2. Kinetic energy should be conserved with slightly more error than for momentum because velocity is squared.

(continued)

Teacher Notes (continued)

3. If energy and momentum are conserved within error, the collision is elastic.
4. Friction from the carts hitting the track during collision is probably the largest source of error other than measurement error.
5. Energy can be lost to friction during the collision. Energy should not be gained. Most likely, energy gain is associated with a measurement or calculation error.

LAB QUIZ ANSWERS

1. c
2. b
3. d
4. b
5. a

Lab #13
A Collision in Two Dimensions

Name: ______________________

Date: ______________________

In this lab you will investigate collisions in two dimensions. Two balls are used. The first ball rolls down an incline and then hits another. The collision causes the balls to move at different angles. The horizontal distance which each flies will be measured as well as the direction of the flight. These data will allow you to see if momentum has been conserved.

Materials

clamp	string
ramp	washer
tape	pencil
paper	meter stick
2 balls of equal mass	protractor
carbon paper	

Procedure

1. Clamp the ramp so that the lower end is just off the edge of the table. Make sure the ramp is horizontal at the table edge. See Figure 13.
2. Take some string long enough to reach from the end of your ramp to the floor and tie a washer on it. Hold the string at the end of the ramp so that the washer is just off the floor. Gravity makes the washer hang vertically and you can find the point right under your ramp. Tape a piece of paper to the floor under the washer and then mark this point with an *X*.
3. Place a ball at the highest point and release it. Note approximately where it lands.
4. Tape a piece of white paper to the floor with its center approximately where the ball landed. Then place a piece of carbon paper over the white paper.
5. Roll the ball down the ramp from the top. When it hits the carbon paper it leaves a dot on the white paper. Roll the ball down the ramp a few more times.
6. With a meter stick, measure the distance from each of your dots to the *X*. Record these values.
7. Draw a line on the paper from the *X* on your paper under the ramp toward the middle of the black dots where the ball landed.

(continued)

Lab #13
A Collision in Two Dimensions *(continued)*

Name: ____________________

Date: ____________________

Figure 13

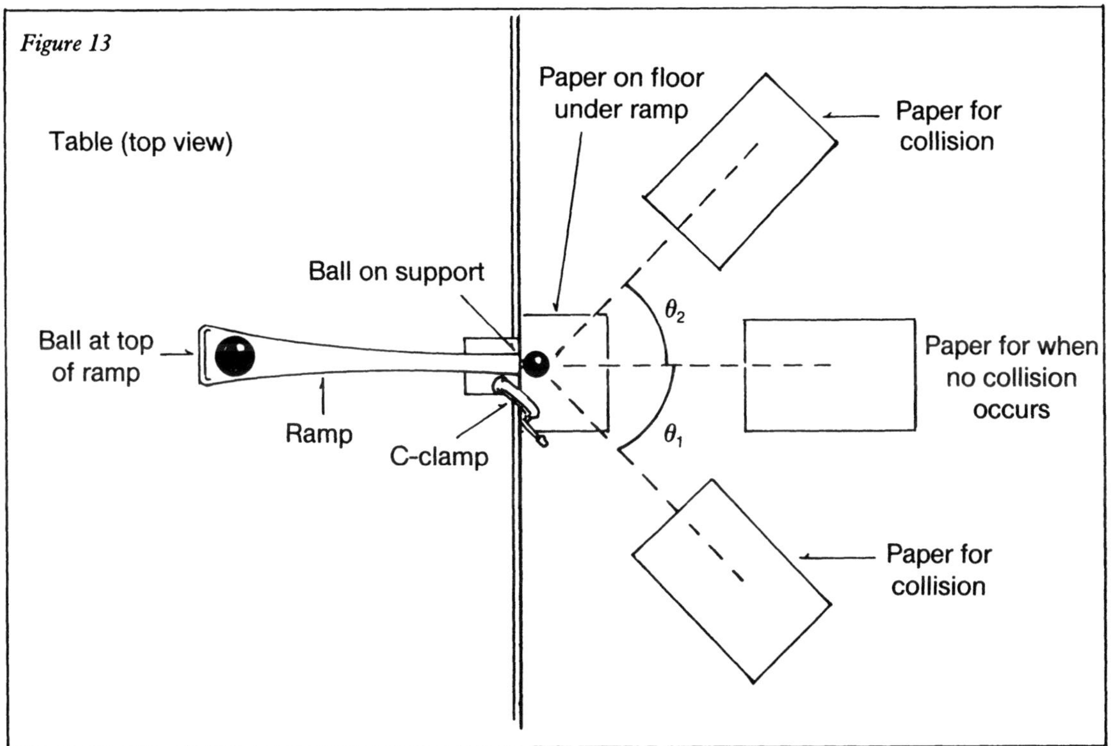

8. Now move the support at the lower end of the ramp so that the moving ball will hit a ball placed on the support. Put a ball on the support and roll the other ball down the ramp and observe the collisions. You do not want a collision where the moving ball barely hits the stationary ball or a head-on collision. Both balls should also move horizontally initially after the collision.
9. Once you have the balls colliding properly, tape a piece of paper to where each ball lands. Put carbon paper on top of each piece of paper. Have the balls collide five times, each time hitting their respective papers.
10. Remove the carbon paper and measure the distance to each dot. Record these data.
11. On the paper under the ramp, draw a line from the *X* toward each of the two clusters of dots. Find the angle between the line you drew toward each cluster and the line to where the single ball landed when it did not collide.
12. Finally, measure the height of the ramp above the floor.
13. Remove your pieces of paper from the floor and keep the one with the *X* and the angles on it.

(continued)

Lab #13

A Collision in Two Dimensions
(continued)

Name: ______________________

Date: ______________________

Cautions

1. Make sure that the balls move horizontally after leaving the ramp.
2. Do not remove your pieces of paper taped to the floor until you have all the proper measurements.

Analysis

1. Find the average distance that each set of black marks was from the position right under the ramp. Give some estimate of the error in this value.
2. Using the height of the ramp above the floor, find the time for the ball to fall from the ramp to the floor. Assume that the vertical velocity is initially zero.
3. Each ball moved horizontally the distance you found in step 1 in the time you found in step 2. Calculate each ball's velocity when it left the ramp. For the two balls which collided, give the angle at which they went. Write these on the piece of paper with the angles and the X on it.
4. Turn the paper so that the velocity vector for the ball which did not hit another ball is pointed to the top of the page. Call this the Y direction. At right angles to this direction is the X direction just as on a graph.
5. Find the X and Y components of velocity for each ball which collided. Make sure that you label them.
6. Multiply each of these velocities by m to find the momentum of each ball. (The actual mass does not matter because the mass of each ball is the same.)
7. Add together the momentum in the Y direction for the colliding balls.
8. Add together the momentum in the X direction for the colliding balls but remember that if they are in opposite directions, then you must use a negative sign for one value.
9. Next calculate the kinetic energy ($KE = (1/2)mv^2$) of each ball.
10. Add together the kinetic energy of the two colliding balls.

(continued)

Lab #13
A Collision in Two Dimension *(continued)*

Name: ______________________

Date: ______________________

Conclusions

1. The momentum you found for the ball that did not collide is equal to the momentum before the collision. The sum of the momenta of the two balls in the *Y* direction should equal this if momentum is going to be conserved. Is momentum conserved in the *Y* direction? Is the difference significant, given your measurement errors?
2. Before the collision, the moving ball had no momentum in the *X* direction so there should be no momentum after the collision. Was there any momentum in the *X* direction after the collision? Does your error associated with measurement error explain the differences?
3. Was the collision elastic, given your measurement errors?

Quiz #13
A Collision in Two Dimensions

Name: ______________________

Date: ______________________

1. When a projectile is fired horizontally across a level surface, it will stay in the air as long as
   a. an object thrown up at the same velocity.
   b. an object thrown down at the same velocity.
   c. an object fired horizontally with half the velocity from the same height.
   d. an object dropped from a height a little bit lower than the projectile was fired.

2. If an object flies horizontally off the end of a ramp and goes 4 meters horizontally in .5 sec, its horizontal velocity in m/sec is
   a. 1.
   b. 2.
   c. 4.
   d. 8.

3. Momentum is the product of
   a. velocity and displacement.
   b. velocity and force.
   c. force and displacement.
   d. velocity and mass.

4. In an elastic collision which of the following should be conserved?
   a. momentum
   b. energy
   c. momentum and energy
   d. neither momentum nor energy

5. A moving ball hits a stationary ball. Initially, the moving ball is going in the *Y* direction (at right angles to the *X* direction). After the collision, the momentum in the *X* direction should equal
   a. zero.
   b. the momentum in the *Y* direction.
   c. a value between zero and the momentum in the *Y* direction.
   d. It can't be determined with the given information.

Lab #13: A Collision in Two Dimensions

Teacher Notes

Cautions

1. Make sure that the balls move horizontally after leaving the ramp.
2. Do not remove your pieces of paper taped to the floor until you have all the proper measurements.

This is a long and involved lab but it does show conservation of momentum and energy in a two-dimensional collision. The error can be as much as 20% or more, however, so be prepared! The calculations and collection of data are quite involved and students must work fast if they only have one period to collect data. With two periods few students will finish the calculations. The lab is good as long as the students don't get lost along the way. If it overwhelms them, it may not be worth the effort.

The "collision in two dimension" apparatus is available from some supply houses and costs $15–$20. It can be used for this lab and a lab using projectiles sent horizontally off the ramp. The apparatus is simple and works relatively well. The height of the support for the stationary ball is critical. After the collision, both balls must be moving horizontally. In order to have this happen, their centers must be at the same height. This may require careful adjustment.

A similar lab can be done on an air table using a strobe photo. It is easier to do although getting enough pictures for each lab group can be trying if you have a large number of students. You can take the photographs beforehand and take one sample photo with the students present so they know how you got the photographs. This is preferable to the ramp experiment but the apparatus is much more expensive. The lab is analyzed in much the same way so this write-up can be modified to be used with the air table experiment.

Possible Problems

1. If the students move their paper before they have taken their measurements, they lose their data.
2. Some students will need to be guided through the calculations or they will become frustrated.
3. The idea that the momentum in the direction at right angles to the initial motion should be zero after the collision may baffle some students. It is almost too

(continued)

obvious for some to see. Also, at that stage in the calculations, some students may be frustrated with the lab and they will not appreciate the fact that zero is the correct answer.

Conclusions

1. The momentum in the *Y* direction equals the mass, *m*, times the velocity when the ball does not collide. The sum of the momenta in the *Y* direction after collision should equal this, too. There will be error, perhaps 20% or more in some cases.
2. The momentum in the *X* direction should equal zero because there was no velocity in this direction before the collision. The sum of the momenta in the *X* direction after collision should equal zero, too. There will be error of similar numerical size to the error in the *Y* momentum.
3. Kinetic energy should be conserved although some error will exist. Therefore, it should be an elastic collision.

LAB QUIZ ANSWERS

1. c
2. d
3. d
4. c
5. a

Lab #14
Ballistic Pendulum

Name: ______________________

Date: ______________________

In this lab you will investigate one method used to find the velocity of rapidly moving objects such as bullets or arrows. An arrow will be shot into a massive target which is suspended by long strings. When the arrow collides with the target, momentum will be conserved. The target will start moving and swing upward until its kinetic energy after the collision has been changed into potential energy. By measuring how high the target swings, you can calculate the velocity of the arrow.

Materials

balance	bow
target	meter stick
arrow	string

Procedure

1. With a balance, find the mass of the target and the arrow. Record these values.
2. Support the target from the ceiling so that it will not turn when it is hit (see Figure 14.1).
3. When the target is not moving, fire the arrow into the target from close range. Make sure that you hit as close to the center as you can.
4. After you have fired the arrow, quickly use a meter stick to measure the distance from where the target is stopped at one end of its swing to where it is stopped at the other end of its swing.
5. Measure the distance from the target's center to the support.

Caution

Do not shoot the arrow so that it might strike anyone. It is dangerous and can cause injury. Make sure that no one is in the area in front of the point where the arrow is released.

(continued)

Lab #14
Ballistic Pendulum *(continued)*

Name: ______________________________

Date: ______________________________

Analysis

The distance the target swings can be used to find the height to which the target rises. The height is found using the equations:

$$h = 1(l\text{–}\cos\theta) \qquad \sin\theta = x/2l$$

where h is the height the target rises, l is the distance from the center of the target to the support, and x is the distance the target swung. Figure 14.2 shows how this is derived.

1. Find the height to which the target rises using the equations.
2. Find the initial velocity of the target for a rise to this height using conservation of energy.
3. Find the momentum of the target after it was struck.
4. Using conservation of momentum, find the velocity of the arrow.

Conclusions

1. Does the velocity at which the arrow is moving seem reasonable? Explain.
2. How large an error do you expect in your answer? What could cause the error?
3. Did the velocity at which the target moved after the collision surprise you?

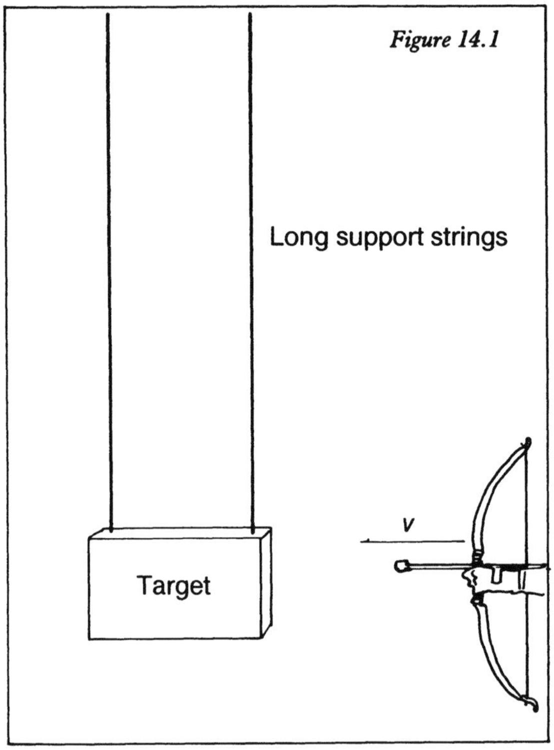

Figure 14.1

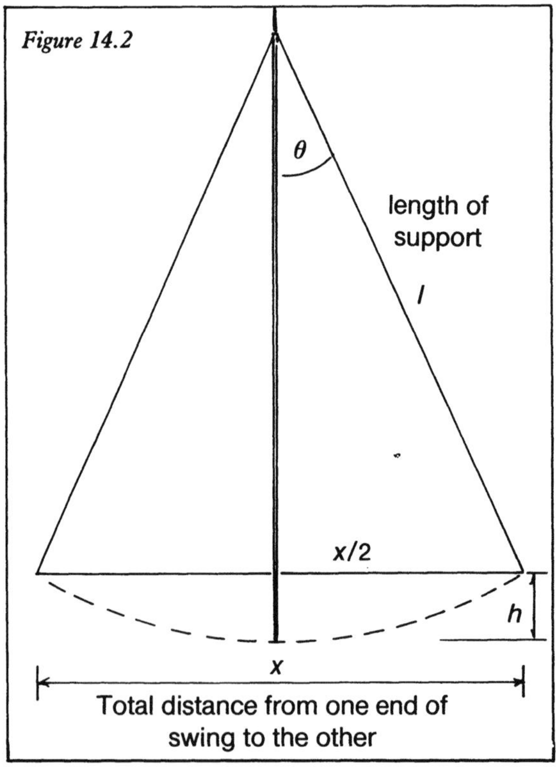

Figure 14.2

Quiz #14
Ballistic Pendulum

Name: ______________________________

Date: ______________________________

1. As the arrow enters the target, what will be conserved?
   a. energy only
   b. energy and momentum
   c. momentum only
   d. neither energy nor momentum

2. When the target swings upward after it is struck by the arrow, which of the following is conserved?
   a. energy only
   b. energy and momentum
   c. momentum only
   d. neither energy nor momentum

3. The momentum of a 2-kg mass moving at 3 m/sec is
   a. 2 kg-m/sec.
   b. 5 kg-m/sec.
   c. 6 kg-m/sec.
   d. 5 m/sec.

4. A 2-kg mass is moving at 6 m/sec. Its kinetic energy in joules is
   a. 8.
   b. 12.
   c. 36.
   d. 64.

5. A 2-kg mass rises 5 meters. What is its increase in potential energy in joules? (g = 10 m/sec^2)
   a. 10
   b. 17
   c. 25
   d. 100

Lab #14: Ballistic Pendulum

Teacher Notes

Caution

Do not shoot the arrow so that it might strike anyone. It is dangerous and can cause injury. Make sure that no one is in the area in front of the point where the arrow is released.

This lab is fun to do and has the students use conservation of momentum and energy to find a result. The lab can be run in many different ways, some giving better results than others and some more dangerous than others. This bow and arrow lab gives good results if you use a standard archery set. If you supervise the actual shooting closely, it will be safe.

A normal archery set such as those used for target practice is suitable for the lab. The only problem is that the target should not have as large a mass as a standard archery target. It may be necessary to build a target for the lab. One method for building the target is to build a box of 1/2-inch-thick plywood which is 12 inches square. One side of the box should be open and the side opposite the open side should have two sheets of plywood for a total thickness of one inch. The box should then be filled with alternating layers of cardboard and straw which will slow down the arrow and also hold it in the target. Test your target to make sure that the arrows do not escape. Replace the cardboard when it has started to lose the ability to hold the arrows.

Tell the students to fire the arrow near the center but not in a hole into which another arrow has gone. The arrow must stay in the target. *Students should fire from a distance of less than half a meter and there should be no one standing in the area behind the target.*

The target should be supported on the longest strings possible. You may be able to use a gym or the lobby of a building. At a minimum, you should have the target a little above the floor and support it from the ceiling.

The apparatus can be purchased from Science Kits #64614 for $200. It produces reliable data with no danger of an arrow hitting a student.

Possible Problems

1. The distance which the target moves is quite small. The measurement of this may be difficult and must be done with care.

(continued)

2. The method of finding the height to which the target rises may confuse students. It is a mathematics problem and confusing if the student is not familiar with sines and cosines.

Conclusions

1. The velocity of the arrow is generally slower than most students think it should be.
2. A reasonable error is 10–20% for the velocity but the actual value can't be determined easily. The largest error is from the measurement of the swing. If this can be measured precisely, then the error can be small.
3. The low velocity of the target is generally surprising to students.

LAB QUIZ ANSWERS

1. c
2. a
3. c
4. c
5. d

Lab #15
Springs and Masses

Name: ______________________

Date: ______________________

In this lab you will investigate how a mass at the end of a spring moves. Many things act as masses bouncing on the end of a spring, including a car, a pogo stick, and a fishing rod. In each case a mass bends a spring and a force tends to make the spring return to its original shape. The period, or the time for the mass to bounce up and down, is determined by the spring constant and the size of the mass. In this lab you will determine the spring constant for a spring and study how the period changes with a change in mass.

Materials

spring

support

meter stick

several masses which can be attached to the spring

timer

Procedure

1. Attach your spring to the support and, with a meter stick, measure the distance from the support to the bottom end of the spring (see Figure 15). Measure to the nearest .001 m (millimeter) and record this distance in a data table similar to the one shown. This is the length of the spring with no mass attached, or its "resting" length.

Data Table

MASS	MG	POSITION	STRETCH	TIME (1)	TIME (2)	AVG. TIME	PERIOD
(kg)	(N)	(m)	(m)	(sec)	(sec)	(sec)	(sec)

2. Attach a .1-kg mass to the spring and with the mass stationary, measure the new distance from the support to the end of the spring. Record the distance and the mass.

(continued)

Name: ____________________

Date: ____________________

Figure 15

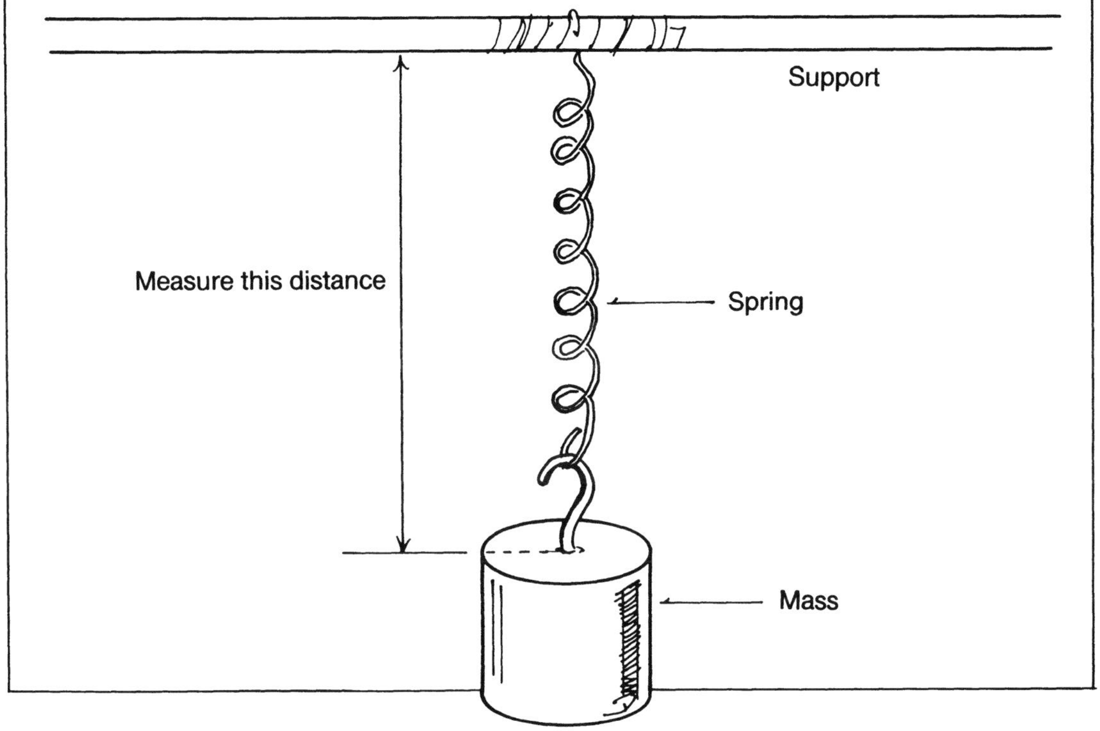

3. Start the .1-kg mass bouncing so it moves about 1 cm up and down. Time how long it takes to make 20 oscillations. Record this time in your data table. Repeat your measurement to make sure that you obtained a good result.
4. Repeat your measurement in step 3, but start the motion by displacing the mass four times as much (4 cm). Does the size of the oscillation make a difference in the time?
5. Repeat your measurements in steps 2 and 3 as you increase the mass by the amounts indicated by your teacher. Do not add more mass than your teacher indicates. Displace the masses only about 1 cm when you make them bounce.

Caution

Make sure that you don't make the mass bounce too far up and down. Such an oscillation can damage your spring by overstretching it and you will then have to start collecting data over again.

(continued)

Lab #15
Springs and Masses *(continued)*

Name: ______________________

Date: ______________________

Analysis

1. Find the force of gravity exerted on each mass ($f = mg$) using $g = 10$ m/sec². Record this in your data table under "*Mg*."
2. Find the distance the spring stretched for each added mass by subtracting the initial length of the spring (with no mass) from the displacement. Do this for each value of mass which you added and record this in the table under "*Stretch*."
3. Draw a graph of force vs. amount stretched (using the values you just found) with the amount stretched on the horizontal axis.
4. Find the spring constant for your spring ($k = f/s$). You can do this either from the slope of the graph or from your data table.
5. Find the average time for 20 oscillations and the time for one oscillation (the period).
6. Draw a graph of mass vs. period for your data using the values you just calculated in step 5.
7. Choose one of the masses and see if the equation for the period of oscillation for a mass on a spring works. The equation is: $T = 2\pi\sqrt{m/k}$ where T is the period in sec, m is the mass in kg, and k is the spring constant (calculated in step 4) in N/m.

Conclusions

1. Does the graph of force vs. stretch show that there is a spring constant for your spring? Does the spring stretch twice as far when twice as much force is applied?
2. What does the graph of mass vs. period look like? Does it agree with the equation for the period of the oscillation? (Should the period increase or decrease with an increase in mass?)
3. Does the period of oscillation as calculated from the equation agree with your measured value? State numbers to support your statement.
4. What sources of error could exist in this lab and how large do you think they are?
5. If you knew the mass of a car (which you can generally find in the owner's manual), how could you find the spring constant of the car's springs?
6. If you cast a heavy lure with a fishing rod, the rod "takes longer" to spring back as compared to when you cast a light lure. Explain.
7. Would different people bounce on a pogo stick at different rates? Explain. Try the experiment if you can.
8. Does the size of the oscillation affect the period of a mass bouncing on a spring?

Quiz #15
Springs and Masses

Name: ______________________

Date: ______________________

1. Spring constant is defined as
   a. force/displacement.
   b. displacement/force.
   c. force/velocity.
   d. displacement/period.
2. Which of the following does not involve a mass bouncing on a spring?
   a. a car
   b. a pogo stick
   c. a pan balance
   d. a lure on a fishing rod
3. The period of oscillation with a mass moving on a spring is the time
   a. for the mass to go completely down.
   b. for the mass to go completely up.
   c. for the mass to go completely down and completely up.
   d. for the mass to move from highest point to its unstretched position.
4. You should avoid having large oscillations because you might
   a. damage your spring and make your data useless.
   b. damage the mass because of large accelerations.
   c. have too large a period and not be able to measure it.
   d. have the frequency too high and not be able to measure it.
5. Atomic bonds act like little springs which hold little masses, the atoms, together. These little atomic springs and atomic masses might be expected to
   a. behave like large masses on large springs.
   b. not behave like large masses on large springs at all.
   c. spin rather than move back and forth or up and down.
   d. behave like no other form of matter.

Lab #15: Springs and Masses

Teacher Notes

Caution

Make sure that you don't make the mass bounce too far up and down. Such an oscillation can damage your spring by overstretching it and you will then have to start collecting data over again.

This lab is a good introduction to springs and it also allows students to work with graphs. Generally, good results are obtained as long as the following spring characteristics are met.

The spring should have a relatively low spring constant, meaning that it should stretch about 1 cm when .1 kg is hung from it (k = 100 N/m). It also must be able to support 2 kg without deforming. (Students should not add more than a kilogram to the spring, but there should be a safety margin or students will deform the springs.) You should test the springs to see the maximum amount of mass which they can take and tell the students this value. Enforce the 1-cm oscillation for any large masses and you will protect your springs.

You may want to do further investigation of the period of oscillation vs. mass. You may want to plot the function of T^2 vs. mass which should give a straight line. Good students may learn from this exercise although many will be mystified by the math.

Possible Problems

1. If the support is not steady, the masses may start to move in circles as they bounce. This does not affect the period but it may distract the students.
2. Students may not understand that it is the amount stretched, not the distance from the support to the string, which is important.

Conclusions

1. The graph is a straight line so it shows $f = -ks$.
2. The graph is a curved line which agrees with the idea that T is related to the square root of m.
3. The period of oscillation should agree with the equation, given that the students have not made calculation or large measurement errors.

(continued)

Teacher Notes (continued)

4. The largest source of error is measuring displacement unless the spring is deformed by stretching it past its elastic limit.
5. Make the car bounce on the springs, find the period of oscillation, and then use the equation.
6. More mass makes the period longer, given the equation.
7. More mass makes the period longer, given the equation.
8. The size of oscillation should not affect the period unless the elastic limit is exceeded.

LAB QUIZ ANSWERS

1. a
2. c
3. c
4. a
5. a

Lab #16
Inertial Pendulum

Name: ____________________

Date: ____________________

In this lab you will investigate a device called an inertial pendulum. The inertial pendulum has two pieces of spring steel which cause the mass to return to an equilibrium position (a place where the net force is zero). With this device, it is possible to remove the gravitational force from a mass and see if that force, the weight of the object, has an effect on the period.

Materials

clamps	masses
an inertial pendulum	timer
balance	

Procedure

1. Clamp one end of the inertial pendulum to the lab bench so that it is firmly attached (see Figure 16). Move the free end to the side and release it. The free end will move back and forth rapidly. Tighten the clamps if the pendulum moves.
2. Using a pan balance, find the mass of the other clamp. Record this value.
3. Clamp the other clamp to the free end of the inertial pendulum. Then time how long it takes for the pendulum to make 20 cycles. (A cycle is defined as going back and forth or the time it takes for the motion to start repeating.) Record this time and the mass attached to the pendulum in a data table. Time at least twice.
4. Clamp .1 kg of mass to the free end of the pendulum. Time at least twice to see how long it takes to make 20 cycles. Record the total mass (clamp and mass) and the times.
5. Keep adding mass in .1-kg units until you have a total mass of about 1 kg. If it looks like the pendulum is about to bend and break, check with your teacher before timing for 20 cycles or adding more mass. Record these times and masses in a data table.
6. Remove all the mass except the clamp and then put an unknown mass through the hole in the free end. Support it with equal amounts showing above and below the hole. This will help prevent it from wobbling. Then time at least twice to find the time for 20 cycles. Record these times in a data table. Label the mass "unknown."

(continued)

Name: ______________________

Date: ______________________

Figure 16

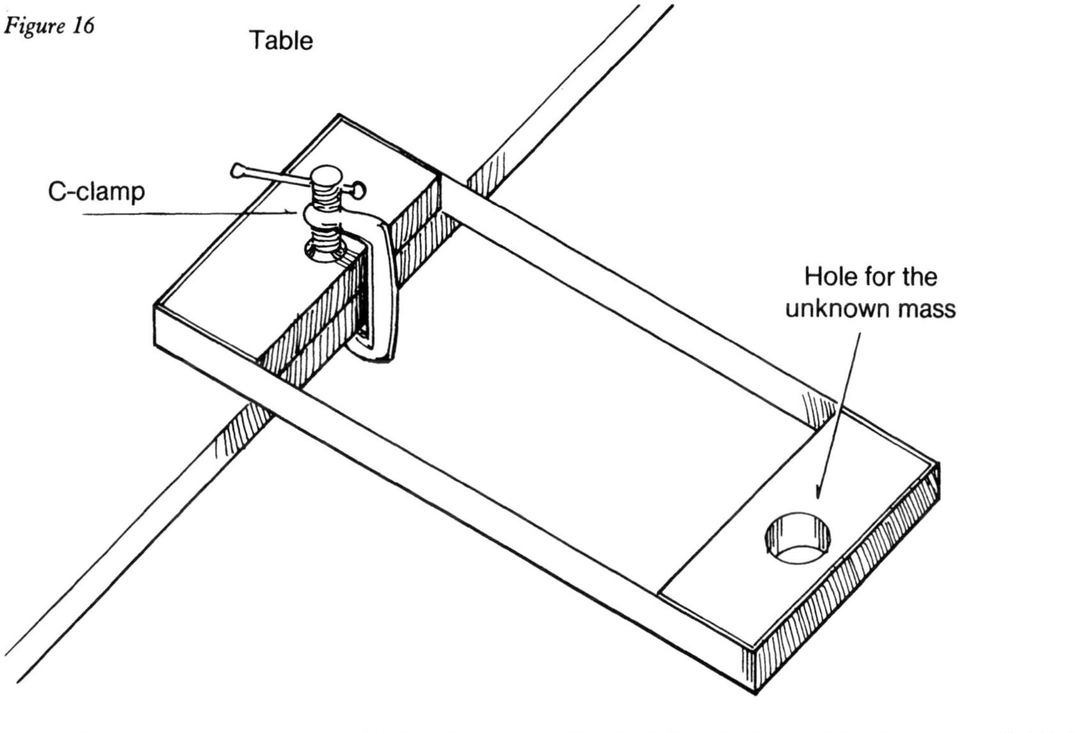

7. Remove the support and attach a string to the unknown mass so that you can support it by holding the string. In this way, you will remove the weight from the pendulum (you support the weight), but the pendulum still must accelerate the mass back and forth. The effect of mass is still present. Time at least twice to find the time for 20 cycles. Record these times in a data table. Label the mass "unknown/no weight."

Caution

Be careful with the timing for it will affect your ability to find the value of the unknown mass.

Analysis

1. Find the average time for 20 cycles for each different mass you used on the pendulum. Record the time in your data table.

(continued)

Name: ______________________

Date: ______________________

2. Find the time for one cycle. This is the period. Record the period in your data table.
3. Plot a graph of your data for mass vs. period with period on the horizontal axis. Draw a smooth curve through the points. Do not connect the dots. Leave some of the points above your line and some below. If it looks like it should be a curve rather than a straight line, draw it as a curve.
4. Using your graph, find a value for the mass of the unknown.

Conclusions

1. Use your graph to explain what happens to the period of the inertial pendulum as the mass increases.
2. Knowing what you do about the equation $f = ma$, why might the period be related to the mass as shown by the graph?
3. What is the unknown mass? What happens when the weight is removed? Is there a noticeable change? Does weight seem to affect the period of motion?

Quiz #16
Inertial Pendulum

Name: ____________________

Date: ____________________

1. You are going to find the mass of an unknown mass and also determine if the ______________ of the mass affects the period of the pendulum.

   a. shape
   b. material
   c. weight
   d. color

2. You are going to draw a graph of ______________ and use it to find the unknown mass.

   a. weight vs. mass
   b. mass vs. force
   c. mass vs. period
   d. period vs. weight

3. If the pendulum swings back and forth 20 times in 15 seconds, what is the period of the pendulum in seconds?

   a. .25
   b. .50
   c. .75
   d. 1.33

4. The equilibrium position is a place where

   a. all things are equal.
   b. all forces are balanced.
   c. most things are equal.
   d. most forces are balanced.

5. You must be most careful about making which particular measurements in this lab?

   a. weight
   b. mass
   c. period
   d. time

Lab #16: Inertial Pendulum

Teacher Notes

Caution

Be careful with the timing for it will affect your ability to find the value of the unknown mass.

In this lab students look at the effect of mass vs. weight. Students also practice graphing and use a graph to find an unknown. It is quite a simple lab which generally produces good results.

The inertial pendulum is available from supply companies for about $25 but it is also possible to build one inexpensively. Two small pieces of wood, two hacksaw blades, a metal bar, some screws, and a drill are all that are needed for materials. The drill is used to make holes in the hacksaw blade and the large hole where the unknown mass is suspended. The hacksaw is used to cut the bar into appropriate lengths.

An unknown mass is supplied with the inertial pendulum materials from supply companies, but can be made from any piece of metal bar. It is desirable not to have the mass of the unknown close to .1, .2, .3, .4 kg, etc. If it is, students will not have to use their graph to find the value; they can just look at their data table. The unknown mass is supported in a hole drilled in the wood to a diameter slightly larger than the metal bar. It can be supported using tape wrapped a number of times around the mass. The tape can secure the string which students hold when supporting the mass.

If you have the students find the mass of their clamps on the pan balance, you will have to remove the balances before you give out the unknown masses. The alternative is to mark the mass on each clamp.

Possible Problems

1. Timing is very important and students may not take enough care or may time improperly. They must start the pendulum and allow it to move back and forth before they count "one." They must not count "one" when they release it.
2. Interpolation from a graph may give some students trouble.
3. Drawing the smooth curve will give some trouble because they will want to connect the dots. In all parts of the graph, some points should be above and some should be below the line.

(continued)

Teacher Notes (continued)

Conclusions

1. As the mass increases, the period increases. Because it is not a straight line, when you double the mass you do not double the period.
2. With more mass, the acceleration is less (because the force is constant) so it takes a longer time to reach the maximum velocity and a longer time to stop. This means that the period is longer.
3. The unknown mass should have the same period whether it is supported by a student or by the pendulum. This follows from the equation $f = ma$ and the discussion for conclusion 2.

LAB QUIZ ANSWERS

1. c
2. c
3. c
4. b
5. d

Lab #17
Centripetal Force

Name: ____________________

Date: ____________________

In this lab you will investigate what factors affect the centripetal force which makes an object move in a circle. The equation for centripetal force, $f = mv^2/r$, indicates that mass, velocity, and radius affect the force. Each of these three quantities will be examined.

Materials	
balance	meter stick
rubber stoppers	pen
string	timer
tube with smooth ends	

Procedure

1. Use a balance to find the mass of a stopper. Record the mass.
2. Thread the string through the tube and then tie a rubber stopper on each end.
3. Use a pen to mark a distance of .5 m along the string from one of the stoppers. This stopper will be the one which moves in a circle. The other stopper will hang down below the tube. See Figure 17.
4. Hold the tube vertically and swing the stopper around in a circle. Practice until you can keep the pen mark at the top of the tube. This will make the stopper move in a circle of .5 m. Avoid moving your hand in a large circle because that increases the radius in which the stopper moves.

Case 1:

5. Time how long it takes for the swinging stopper to make 20 complete revolutions and record the time. Repeat this measurement more than once to see if you have the correct time. Keep the mark you made on the string at the top of the tube for the entire time.

Case 2:

6. Pull some of the string down through the tube so there is .25 m between the stopper and the top of the tube. Put another mark on the string so that you can keep the stopper swinging at this position. Again find the time for 20 revolutions when the centripetal force is just balanced by the gravitational force of the hanging stopper. Record the time, number of stoppers, and radius. Repeat this measurement more than once to see if you have the correct time.

(continued)

Name: ______________________

Date: ______________________

Figure 17

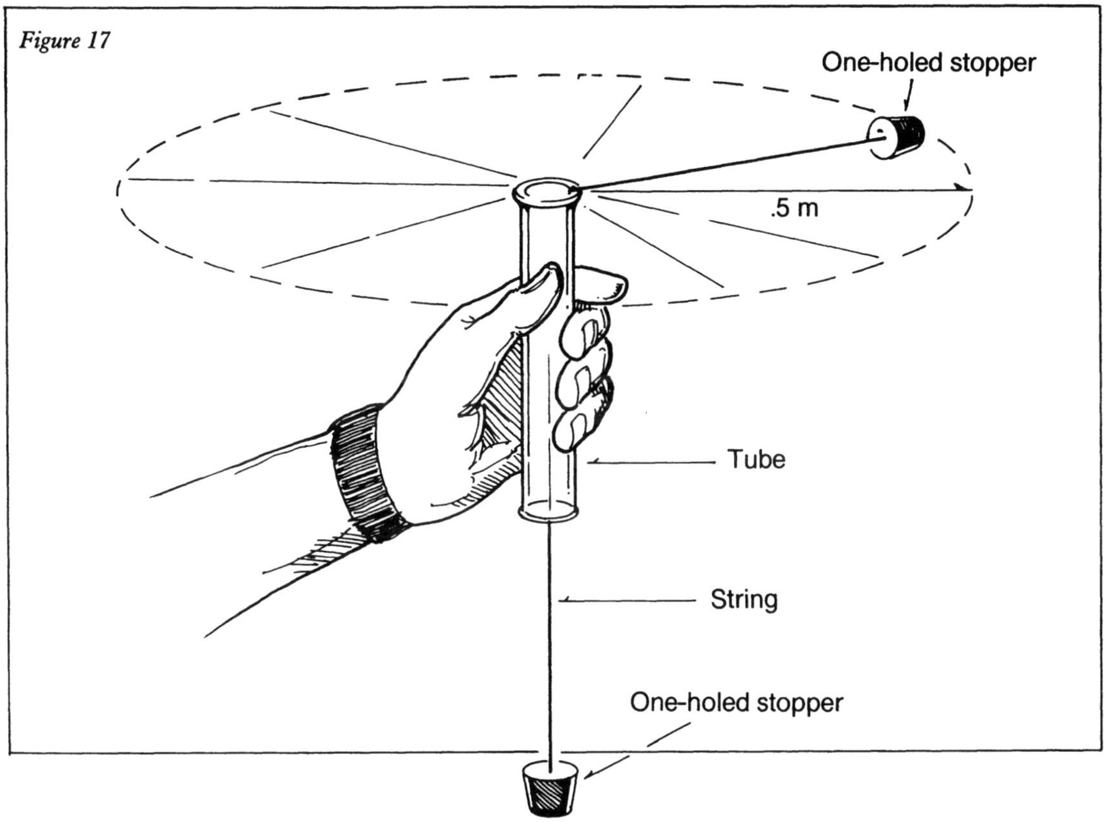

Case 3:

7. Tie three more stoppers to the hanging one so that there are four hanging stoppers. Use the first radius (about .5 m) and find the time for 20 revolutions when the centripetal force is just balanced by the gravitational force of the hanging stoppers. Record the time, number of stoppers, and radius. Repeat this measurement more than once.

Case 4:

8. Remove two of the hanging stoppers so that there are two remaining. Tie a second stopper to the swinging end of the string. Make sure that the radius is still .5 m when you tie on the extra stopper. (You may have to make a new mark on the string.) Time for 20 revolutions when the centripetal force is just balanced by the gravitational force. Record the number of stoppers, radius, and time. Repeat this measurement more than once to see if you have the correct time.

(continued)

Lab #17
Centripetal Force *(continued)*

Name: ______________________

Date: ______________________

> ***Caution***
>
> The swinging stoppers are going fast enough to hurt someone. Make sure that you have enough room to swing them without hitting anyone.

Analysis

1. Find the average time for 20 revolutions in each of the four cases.
2. Find the time for one revolution in each of the four cases.
3. Using radius and the equation for the circumference of a circle ($2\pi r$), find the distance the swinging stopper goes in one revolution.
4. Using the time and distance for one revolution, find the velocity at which the stopper is moving.
5. Calculate the centripetal force using the equation for each case.
6. Calculate the gravitational force on the hanging mass in each case, given the mass of each stopper and the number of stoppers.

Conclusions

1. In Case 1, does the centripetal force equal the gravitational force? How much error is there? Is this amount of error reasonable?
2. In Case 2, the radius was made half as large. Did the velocity change as you would have expected from looking at the equation? Does the centripetal force still equal the gravitational force within error?
3. In Case 3, the radius was returned to the original value but the gravitational force was made four times larger. Did the velocity change as you would have expected from looking at the equation? Does the centripetal force still equal the gravitational force within error?
4. In Case 4, the swinging mass and the hanging mass were doubled as compared to Case 1. Did the velocity change as you would have expected from looking at the equation? Does the centripetal force still equal the gravitational force within error?
5. Does this experiment help to support the equation for centripetal force given that there might be 10–20% error? What could be the source of this error? Could the sources of error you suggest reasonably explain the error which exists?
6. When you swing the mass in a circle, your hand most likely moved in a small circle. What effect does this have on the results?

Quiz #17
Centripetal Force

Name: ______________________

Date: ______________________

1. In this lab the centripetal force is balanced by a

   a. gravitational force.
   b. pan balance.
   c. pull on a string by a hand.
   d. spring scale.

2. If a mass is moving in a circle of radius 1.00 meter and it takes 1.00 second for it to complete the circle, what is its velocity in m/sec?

   a. .5
   b. 2
   c. 3.14
   d. 6.28

3. Using the equation for centripetal force, if you double the mass moving in a circle (all other things remaining constant), the centripetal force

   a. is halved.
   b. remains the same.
   c. is doubled.
   d. is four times greater.

4. Using the equation for centripetal force, if the velocity at which the mass moves is doubled (all other things remaining constant), the centripetal force

   a. is halved.
   b. remains the same.
   c. is doubled.
   d. is four times greater.

5. Using the equation for centripetal force, if the radius in which a mass moves is doubled (all other things remaining constant), the centripetal force

   a. is halved.
   b. remains the same.
   c. is doubled.
   d. is four times greater.

Lab #17: Centripetal Force

Teacher Notes

> ***Caution***
>
> The swinging stoppers are going fast enough to hurt someone. Make sure that you have enough room to swing them without hitting anyone.

This lab helps students to understand and actually "feel" the changes in centripetal force. Friction and student error may cause some problems with accuracy. The frictional problems are caused by the string moving at the top of the tube. The student errors are caused by students moving their hands in large circles (increasing the radius) or not keeping the marker at the proper place.

The tubes are critical in this lab. They should be large enough to fit in the students' hands and strong enough not to break. Their tops must be made so that they allow the string to swing with as little friction as possible; otherwise friction becomes a problem and the string may be cut. Copper pipe can be smoothed somewhat with sandpaper although it will always tend to be sharp.

If you have some skills with cutting glass, a large test tube can be fashioned into a good tube. The upper lip is smooth and gives good results. Cut off the bottom, and heat the glass until it loses its sharp edges. This requires relatively high temperatures if you are using Pyrex® glass. Then wrap the glass with a number of layers of electrical tape to prevent it from shattering if broken. Warn students to treat these wrapped test tubes carefully.

A more expensive alternative and one which has a little more friction is to use the attachment fitting used to connect the water lines to a toilet or sink. This fixture is a tube with a Teflon® collar on one end. The advantage of this device is that it will not break into sharp pieces if it is treated roughly.

The masses should be one-holed rubber stoppers. They can easily be tied on the end of a string, are hard to damage, do not hurt as much as metal if they hit someone, and are less likely to cause damage if they fly across the room.

Possible Problems

1. Student may not time or measure carefully or may allow the radius to change.
2. Excessive motion of the swinging hand will change the radius greatly. This changes the required velocity. Remind the students to move the tube as little as possible.

(continued)

Teacher Notes (continued)

3. Some students may use only r to find the velocity without using the circumference of the circle.
4. Some students may endanger other students by wildly swinging their masses.

Conclusions

1. The centripetal force should equal the gravitational force plus or minus 10–20%.
2. With half as large a radius, the quantity v^2 must be half as large. This means that the velocity must be smaller by a factor of the square root of one-half. It should be about .71 of the previous velocity. The equation should still work.
3. With four times the gravitational force, the velocity must be doubled in order to have four times the centripetal force. The equation should still work.
4. Because both the masses are doubled, the centripetal and gravitational forces are doubled when the velocity remains the same. Thee equation should still work.
5. Friction of the string on the tube is the main unavoidable problem. A second error source is changing the radius the mass is swung.
6. If the hand moves in a circle, it increases the radius. This question is included because it may get the students to accept more error in their answers. In fact, this probably is the main source of error.

LAB QUIZ ANSWERS

1. a
2. d
3. c
4. d
5. a

Lab #18

Inclined Planes: Why It Is Hard to Go up a Steep Hill

Name: ______________________

Date: ______________________

In this lab you will investigate how the force parallel to a hill increases as the angle increases. This parallel force is a component of, or part of, the gravitational force. You must oppose the parallel force if you are to prevent yourself from sliding down a hill. As the parallel force increases, you have to exert more force. This effect is evident when you go up a hill on a bicycle.

Materials

pulley

ramp

string

protractor

cart with a way to attach extra mass

spring scale

Procedure

1. Attach your pulley to the ramp, tie one end of the string to the cart, and tie a loop in the other.
2. When the ramp is level (at zero degrees), there is no parallel component of the gravitational force. Gravity in that case points in a direction perpendicular to the ramp. In a data table record this value next to zero degrees.
3. Raise your ramp so it is at 10 degrees. Use a protractor to measure the angle and record the force shown on the spring scale (see Figure 18). Repeat the measurement as you increase the angle in increments of 10 degrees until you reach 90 degrees. Record these values in your data table.
4. Add mass to your cart and repeat the measurements every 10 degrees from 0 to 90 degrees.
5. Finally, use a block instead of a cart with wheels. Find the force as you vary the angle in 10-degree increments.

(continued)

Lab #18

Inclined Planes: Why It Is Hard to Go up a Steep Hill *(continued)*

Name: ______________________

Date: ______________________

Figure 18

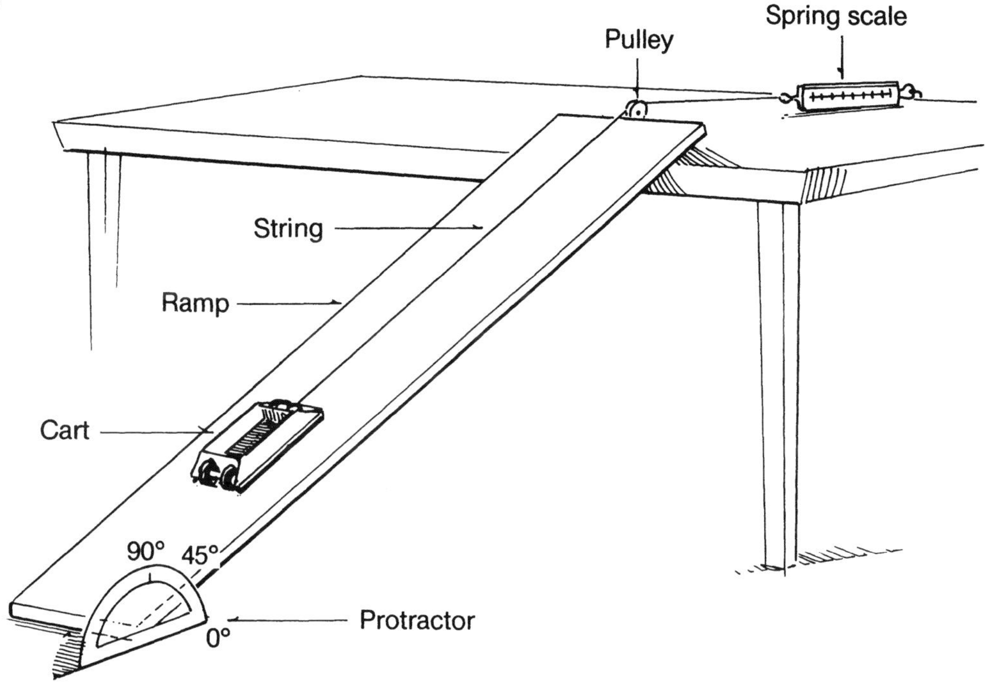

Analysis

1. For the cart with no mass on it, draw a graph of force vs. angle with angle on the horizontal axis. Use *X*s to represent these points. Draw a smooth curve through these points. Some of your points will be above your curve and some below.
2. On the same set of axes, draw a graph of the force vs. angle for the cart when the extra mass was added. Use *O*s for these points. Draw a smooth curve through them.
3. Again, on the same set of axes, draw a graph of force vs. angle for the block. Use *X*s for these points and draw a smooth curve through them.

Conclusions

1. What kind of curve do your graphs look like?

(continued)

Lab #18

Inclined Planes: Why It is Hard to Go up a Steep Hill *(continued)*

Name: ____________________

Date: ____________________

2. What was the effect of adding more mass to the cart? How did it change the graph?
3. What does the graph for the block look like? Why is it different from a cart? Explain.
4. What do these graphs say about your problem of getting up a hill on a bicycle as the hill gets steeper? What sort of force parallel to the hill would you have to exert if you were to go up a vertical cliff? Do you think that this would be possible with a regular bike?

Quiz #18

Name: ______________________

Date: ______________________

Inclined Planes: Why It Is Hard To Go up a Steep Hill

1. As the angle of a hill increases, the force required to keep you from slipping down
   a. increases.
   b. increases and then decreases.
   c. decreases.
   d. decreases and then increases.

2. In this lab you will measure the forces exerted on a cart and the
   a. size of the cart.
   b. length of the ramp.
   c. angle of the ramp.
   d. time for the cart to move 1 m.

3. The gravitational force on the cart on the ramp will always point
   a. horizontally.
   b. at a 45-degree angle from the horizontal.
   c. vertically.
   d. It can't always be determined.

4. When you are drawing your graph, you are supposed to use *X*s when plotting values for the cart alone. The *O*s are used to allow you to see the difference between those values and the values for
   a. a different mass.
   b. a different length.
   c. a different set of angles.
   d. the vertical force.

5. When you walk up a hill, the frictional force opposes the
   a. gravitational force.
   b. parallel component of the gravitational force.
   c. normal component of the gravitational force.
   d. normal component of the frictional force.

Lab #18: Inclined Planes: Why It Is Hard to Go up a Steep Hill

Teacher Notes

In this short lab students become more familiar with the way that the parallel force varies with angle. Drawing a graph may make them more aware of the relationship of math to physics. The lab should also encourage them to think more about force components as they ride or walk up and down hills.

The pulley needs to be firmly attached to the ramp. It is hard to get good data if it wobbles or twists. The spring scales must be able to measure forces equal to the gravitational force of the cart and extra mass.

Possible Problems

1. If students do not measure the angles as they are measuring the forces, they will get poor data. The graph should still look like a sine curve but it will have more scatter.
2. If the spring scales stick, students may not get good data.
3. Students may never have seen sine curves in math.

Conclusions

1. The graphs should look like sine curves. Some students may not have seen sine curves in math classes yet.
2. With more mass, the forces are larger. The shape of the curve is still the same.
3. The block has no force until the parallel component equals the frictional force. Then the force exerted by the the spring increases in a similar manner to the other two curves. A few students might note that as the angle increases, the normal force will decrease so the frictional force decreases. At 90 degrees, the frictional force equals zero.
4. As the hill gets steeper, the parallel force increases, reaching your gravitational force when the angle equals 90 degrees. Frictional forces require a normal force so a regular bicycle could not climb a vertical cliff.

LAB QUIZ ANSWERS

1. a
2. c
3. c
4. a
5. b

Lab #20

Forces on Beams, Torques, and Diving Boards

Name: ______________________

Date: ______________________

In this lab you will investigate how forces vary as weight moves on a beam and how the angle at which a force is exerted relates to the size of the torque. You will also learn why large forces are exerted on the supports of a diving board as compared to the force on the diver.

Materials

spring scales	masses
support	protractor
meter stick	

Procedure

1. Attach two spring scales to the horizontal support one meter apart and then place a meter stick on the spring scales, as in Figure 20.1. (A hanging mass will be added in step 3, below.)
2. Record the force on the left and right spring scale in a data table similar to the one shown on the next page. Record the mass (attached to the meter stick) as zero and the spring scales as at 90 degrees to the meter stick.
3. Place the 1-kg mass next to the left hook, record its position as 0.0 m, and then record the forces on each spring scale. The scales should still be at 90 degrees to the meter stick.

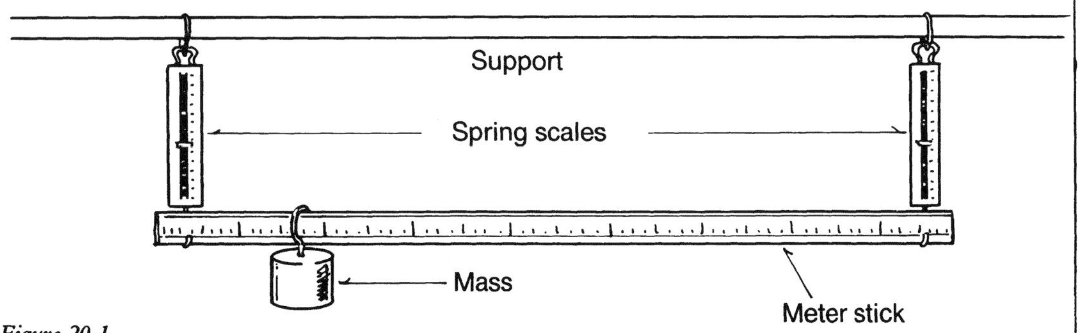

Figure 20.1

(continued)

Name: ______________________________

Date: ______________________________

Data Table

FORCE RIGHT	FORCE LEFT	MASS	DISTANCE FROM LEFT END	ANGLE
(N)	(N)	(kg)	(m)	(deg)

4. Repeat your measurements as you move the mass .1 m at a time along the meter stick until it is next to the right spring scale.
5. Attach .5 kg at the center and record the forces on each spring scale.
6. Leaving the .5 kg at the center of the meter stick, move the supports so that they cause the spring scales to be at an angle of 80 degrees. Use a protractor to carefully measure this angle. It may take a while to get the angle right. You might try holding one spring scale and not attaching it to the support. When you have the correct angle, record the force on each spring scale.
7. Repeat the previous step for angles of 70, 60, 50, 40, 30, and 20 degrees.
8. Finally, move one of the supports so the spring scale is in the middle, .5 m from the left end as shown in Figure 20.2. Then place a .5-kg mass at the right end. You will have to hold the other end because the spring scale will now pull down instead of up. Measure the forces exerted by the spring scales. This setup is similar to the way most diving boards are constructed.

Caution

Make sure that you measure the angles carefully and that you don't change them before you get your measurements of force.

Analysis

1. Use the data from when the spring scales were at 90 degrees to draw a graph of force on the left scale vs. position. Position is on the horizontal axis. Use *X*s to mark these points.

(continued)

Name: ____________________

Date: ____________________

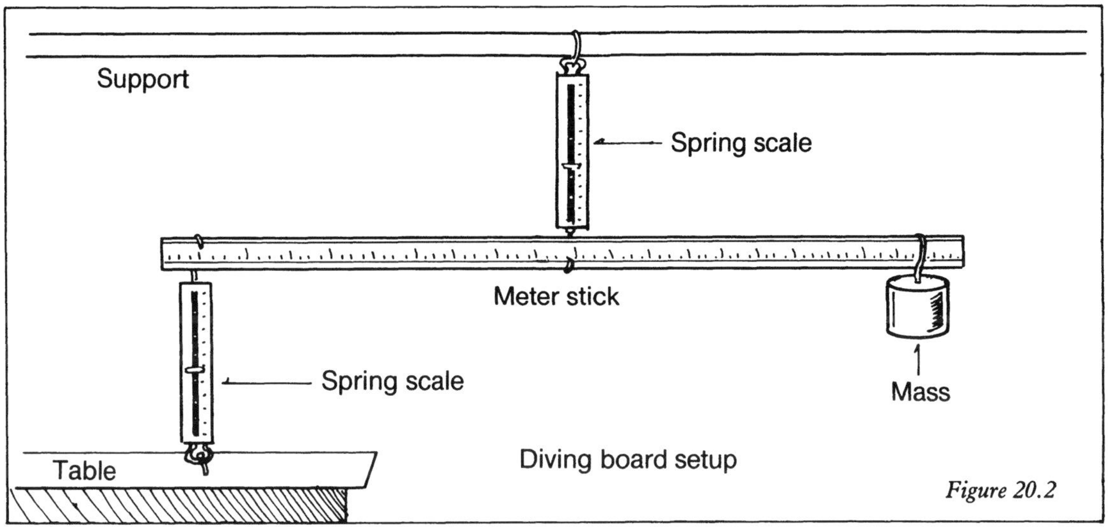

Figure 20.2

2. On the *same* set of axes, draw another graph of the force on the right scale vs. position. Use *O*s to mark these points.
3. Draw smooth curves through the points and label the graphs.
4. Draw another graph of the force on the scales vs. angle. Draw a line through these points.

Conclusions

1. Using the first graph you drew, discuss how the force on each spring scale changes as the mass moves from one end of the meter stick to the other.
2. Using torques, explain why the forces change as the mass moves.
3. When the mass is at the right end, do the readings on the scales make sense?
4. Using the second graph you drew, discuss how the force on each spring scale changes as the angle decreases from 90 degrees. Why does the change occur? What would the force be if the angle were zero degrees? (Is zero degrees possible?)
5. How do the forces exerted on the "diving board" supports (both of the spring scales) compare with the gravitational force exerted by the diver (the .5-kg mass)? Which is largest? Would it be worse for the diver to step on your toe or to have your toe under the diving board? Why?
6. Choose the left end of the board for an axis and use torques to explain why the upward force on the board must be greater than the gravitational force from the diver.

Quiz #20

Name: ____________________

Date: ____________________

Forces on Beams, Torques, and Diving Boards

1. A torque is the product of

   a. force times radius.
   b. force divided by position.
   c. position times radius.
   d. velocity times radius.

2. In this lab a mass is placed on a meter stick and then the position of the mass and the readings from ____________ are recorded in the data table.

   a. a balance
   b. a spring scale
   c. balances
   d. spring scales

3. A force can exert the largest torque when it is exerted at ____________ degrees to the radius.

   a. 0
   b. 30
   c. 45
   d. 90

4. A 4-N force is exerted at right angles to a radius, .5 m from the axis. What is the torque in N-m?

   a. 2
   b. 3.5
   c. 4
   d. 4.5

5. A 60 N-m torque can be balanced by an opposite torque produced by a 20-N force exerted at right angles to the radius at a distance of ____________ meters from the axis.

   a. 1
   b. 2
   c. 3
   d. 4

Lab #20: Forces on Beams, Torques, and Diving Boards

Teacher Notes

Caution

Make sure that you measure the angles carefully and that you don't change them before you get your measurements of force.

This lab can help students to understand torques and how forces on a bridge or beam vary as a mass moves along it. The second part of the lab looks at how torque is dependent on the angle at which the force is applied. Only the component of the force at right angles to the meter stick causes torque. If the angle is zero, the force can provide no torque. The third part of the lab looks at a common situation where forces seem to be magnified. However, if the left end of the diving board is chosen as the axis, the upward force from the diving board support must be larger because it has a smaller moment arm than the diver at the end of the board.

This lab works better if you drill small holes in the meter stick or attach hooks to it so the spring scale hooks go through the holes or hooks. You need holes or hooks at each end and in the middle. It is possible to attach the meter stick to the hooks with tape, but the holes probably do less damage to the meter stick than the tape does in the long run. The other alternative is to use sticks of wood a meter long, mark a centimeter scale on them, and then attach cup hooks. The problem of just using meter sticks with no holes or hooks is that the meter sticks will twist in the spring scales. It is also not possible to exert forces at angles other than vertical.

You will note that no attempt has been made to completely analyze the torques in the first part of the lab where the mass is moved along the meter stick. The problem is that the analysis is difficult unless you remove the forces caused by the meter stick alone. If you instruct the students to subtract this value from each of their measurements, the torque can be analyzed quite easily.

Possible Problems

1. If the spring scales do not measure large enough values, you will have to reduce the masses used and perhaps use a smaller piece of wood.
2. Students often don't like working with torques and need to have someone suggest an axis about which the torques are exerted. Any axis will do, but for ease, you can tell them to use the left end as the axis of rotation. Using an end where a force is exerted makes the analysis much easier.

(continued)

Teacher Notes (continued)

Conclusions

1. Both graphs are similar and when the mass is farther from the support, the support has to exert less force.
2. If the left end is used as the axis, as the mass moves to the left, the clockwise torque increases with distance so the counterclockwise torque provided by the right scale must increase.
3. When the mass is at the right end, the left measures just what it did with no mass. This is just like having the mass hanging on the right scale.
4. As the angle becomes less, the force becomes larger. It is twice as large at 30 degrees and would reach infinity at 0 degrees.
5. The forces are larger than the gravitational force on the diver. It would be worse to have your toe under the diving board because of a larger force.
6. The gravitational force from the diver is exerted at a greater distance from the axis as compared to the support. Therefore, for the torques to be equal, the upward force must be larger than the gravitational force.

LAB QUIZ ANSWERS

1. a
2. d
3. d
4. a
5. c

Lab #21
Torsional Pendulum

Name: ______________________

Date: ______________________

In this lab you will investigate the torsional constant and torsional motion. The torsional constant is like the spring constant but instead of giving the force for a given displacement, it gives the torque for a given amount of twist. You will find a torsional constant and examine how a change in the torsional constant or the moment of inertia affects the period of the torsional pendulum.

Materials

fish line	clamp
boards	meter stick
support	timer

Procedure

1. Tie a fish line through a hole near the center of a piece of wood.
2. Using a clamp, attach the fish line to the support so that the knot which you used to tie on the piece of wood is 30 cm below the support (see Figure 21). Measure this distance precisely. Label this piece of wood *A*.

Case 1:

3. A cycle of the pendulum begins with the pendulum stopped. It swings one way, stops, and then swings back to its original position. Twist it one revolution from its equilibrium position and release it. Find how long it takes for five cycles. Record this time and repeat the measurement.

Case 2:

4. Shorten the fish line so that there are 15 cm between the knot and the support. Again time for five cycles. Record this time and then repeat the measurement.

Case 3:

5. Now tie on another piece of wood, adjust the fish line so that there are 15 cm between the knot and the support, and again time for five cycles. Label the second piece of wood *B*. Record this time and repeat the measurement.
6. Measure the length of each piece of wood and find its mass. Record these values.

(continued)

Name: ______________________

Date: ______________________

> ***Caution***
>
> Make sure that the fish line does not twist in the clamp. This will change the amount of torque that the fish line provides.

Figure 21

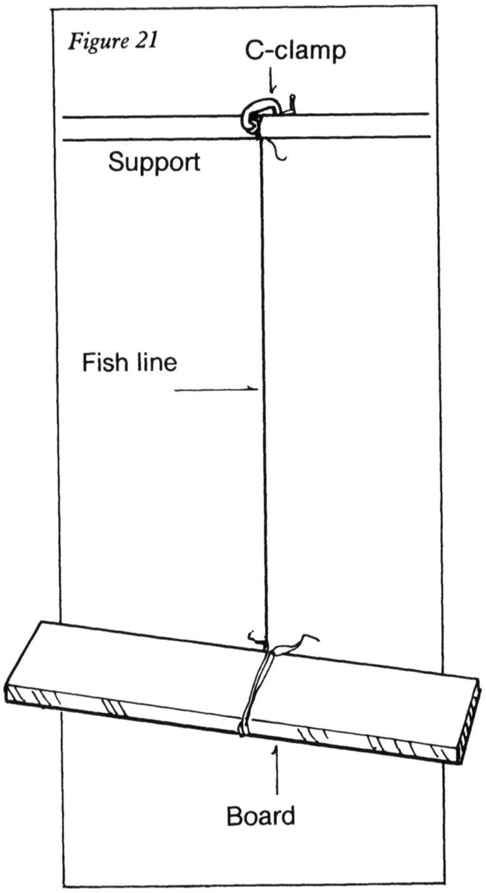

Analysis

1. Find the average time for five cycles for Cases 1, 2, and 3.
2. Find the period of oscillation for each case.
3. Find the moment of inertia for the piece of wood marked *A* using the following equation:

$$I = (1/12)ml^2$$

where I is the moment of inertia, m is the mass, and l is the length of the board.

4. Find the torsional constant which has units of N-m/radian for Case 1 by using the equation:

$$T = 2\pi\sqrt{I/k}$$

where T is the period, I is the moment of inertia, and k is the torsional constant.

5. Find the torsional constant for Case 2.

Conclusions

1. In Case 1 the fish line was 30 cm, and in Case 2 it was 15 cm. What happened to the torsional constant when the length was cut in half? What do you think would happen if the fish line were only 7.5 cm long?
2. If a spring is cut in half, its spring constant is doubled. If a particular spring is stretched .2 m, it gives a force of 10 N. When the spring is cut in half and again stretched .2 m, because the spring is only half as long, each segment of the spring must stretch twice as much. Because this requires twice the force, the spring constant is doubled. Is this also true for the torsional constant? Explain.
3. How does the change in moment of inertia affect the period of motion? Does it change by the amount that the equation would predict? Explain.
4. In what way is a mobile like a torsional pendulum? In what way is it different?

Quiz #21
Torsional Pendulum

Name: ______________________

Date: ______________________

1. The torsional constant is similar to the
   a. gravitational constant.
   b. metric constant.
   c. spring constant.
   d. radiant constant.
2. A torsional pendulum ______________ back and forth instead of swinging back and forth.
   a. jumps
   b. twists
   c. runs
   d. torques
3. Torque is the product of
   a. force and velocity.
   b. velocity and mass.
   c. mass and velocity.
   d. force and radius.
4. The moment of inertia of an object is determined by its
   a. mass and dimensions.
   b. dimensions and velocity.
   c. dimensions and acceleration.
   d. mass and velocity.
5. When the fish line is twisted it will provide a ______________ to the board at the end of the fish line.
   a. force
   b. torque
   c. mass
   d. angular velocity

Lab #21: Torsional Pendulum

Teacher Notes

Caution

Make sure that the fish line does not twist in the clamp. This will change the amount of torque that the fish line provides.

In this lab students look at a torsional pendulum, which they will find interesting. Some clocks run on this principle and they are fascinating to watch. Students will be familiar with mobiles and other such devices which also behave as torsional pendulums.

Shortening the fish line to half its original length means that each segment of the fish line now must twist twice as far for a given angle. Therefore, there is twice the restoring torque from the fish line. The torsional constant is twice as large.

A more expensive apparatus is available from Central Scientific Company (#75260, about $200), but the apparatus described here gets reasonably good data and is inexpensive enough to build for each lab group to have a set. It does not take long to construct the equipment. You need two pieces of 1 × 3 pine board, 8 inches long. Drill a small hole in each board at the edge over the center of mass. When the board is supported by a fish line which goes through this hole, the board should be level. You can use C-clamps to hold the fish line tightly to the support. Thick fish line works well but you want students to be able to tie it easily. The thinner the fish line, the longer the period of oscillation.

Possible Problems

1. If the period of oscillation is too large, slight air motion may affect the period of the pendulum.
2. Students may not like to use the equations. Many students have trouble finding an unknown which is under a square-root sign.

Conclusions

1. The torsional constant should be twice as much. It would be four times as much.
2. Yes, with half the length, each segment of the fish line must twist twice as far when the mass turns. With twice the twisting, the torque will be twice as much or the torsional constant is twice as large.

(continued)

Teacher Notes (continued)

3. The period should increase by a factor of square root of 2. This agrees with the equation.
4. Mobiles are like torsional pendulums but can be more complex because a number of objects twist on a number of strings.

LAB QUIZ ANSWERS

1. c
2. b
3. d
4. a
5. b

Lab #22
Energy and a Bicycle Wheel

Name: ______________________________

Date: ______________________________

In this lab you will investigate how potential energy is changed into rotational energy. A mass will move downward on the pedals of a bicycle and this change in potential energy will be changed into the rotational kinetic energy of the rear wheel.

Materials

bicycle	chalk
support	masses
clamp	string
table	timer
rope	

Procedure

1. Mount the bicycle so that the rear wheel turns freely using rope, a table, and a C-clamp as shown in Figure 22. Place a mark on the tire with chalk so that you can easily see how rapidly the wheel is turning.

Figure 22

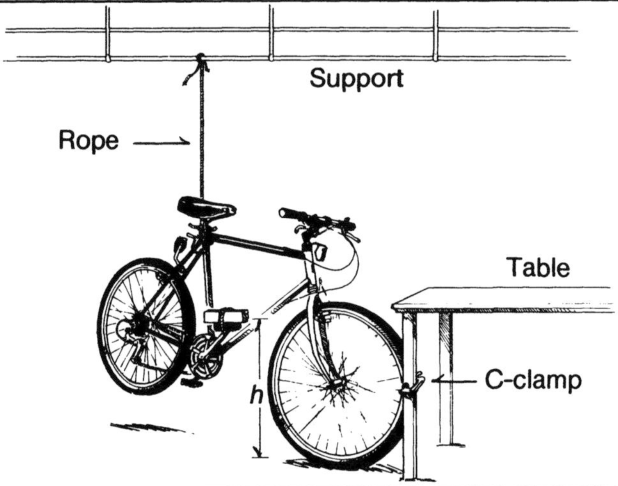

Case 1:

2. Attach 2 kg of mass to a pedal.

(continued)

Name: ______________________

Date: ______________________

3. Move the mass to the highest point where it will still cause the rear wheel to turn when it is released. (If the pedal is vertical, it exerts no torque and will not cause the pedal to turn.) Measure the distance from the highest point to the lowest point as the pedal turns. Record this value.
4. Now move the pedal with the mass on it to the highest point, stop the rear wheel, and then release the pedal so the pedal turns. After it reaches the lowest position, find how many times the wheel turns in 10 seconds. Measure this to the nearest tenth of a turn or more precisely if possible. Record this value and repeat your measurement a few times.

Case 2:

5. Now attach a total of 4 kg to the pedal. Find the number of times the wheel turns in 10 seconds after the mass moves from the highest point to the lowest. Record the results and repeat your measurement a few times.
6. Remove the wheel from the bicycle and find its mass and radius. Record these values.

Cautions

1. Be careful when measuring the number of times the wheel turns. You must be as precise as possible.
2. Make sure the wheel is completely stopped before allowing the mass on the pedal to move down.

Analysis

1. Find the average number of times the wheel turned in 10 seconds for Case 1 and Case 2.
2. Find the number of radians the wheel turns in 10 seconds for each case (2π radians = 1 revolution).
3. Find the angular velocity for each case.
4. Assume that all the mass of the wheel is at the radius. Calculate the moment of inertia of the wheel using the equation: $I = mr^2$.
5. Find the rotational kinetic energy of the wheel in each case ($KE = (1/2)Iw^2$).
6. Find the change in gravitational potential energy in each case (mgh).

(continued)

Lab #22

Energy and a Bicycle Wheel
(continued)

Name: ______________________

Date: ______________________

Conclusions

1. In each case, potential energy should equal the rotational kinetic energy. Which is higher?
2. What sources of error could exist in this lab?
3. Where else could energy go other than to the rotational energy of the rear wheel?
4. What percent of the energy is lost or gained? Do you expect this result? Explain.
5. How did the angular velocity of the wheel when 4 kg moved downward compare with the angular velocity with the 2 kg? Does the equation for rotational kinetic energy predict this?

Quiz #22
Energy and a Bicycle Wheel

Name: ______________________

Date: ______________________

1. In this lab potential energy of a mass moving downward is changed into

   a. rotational kinetic energy.
   b. linear kinetic energy.
   c. spring potential energy.
   d. gravitational potential energy.

2. Mass attached to a pedal of the bicycle is used because it is a source of

   a. rotational kinetic energy.
   b. linear kinetic energy.
   c. spring potential energy.
   d. gravitational potential energy.

3. Rotational kinetic energy is related to the angular velocity and the

   a. mass.
   b. moment of inertia.
   c. force.
   d. radius.

4. A chalk mark is made on the wheel to make it easier to

   a. find the radius.
   b. find the moment of inertia.
   c. measure the time for one revolution.
   d. measure the mass.

5. If the pedal with the mass on it is straight up, the pedal will not move because it provides no

   a. force.
   b. torque.
   c. acceleration.
   d. inertia.

Lab #22: Energy and a Bicycle Wheel

Teacher Notes

Cautions

1. Be careful when measuring the number of times the wheel turns. You must be as precise as possible.
2. Make sure the wheel is completely stopped before allowing the mass on the pedal to move down.

In this lab a bicycle is used to demonstrate the change of potential energy into rotational kinetic energy. Because of friction, the results tend to be only approximate. It is probably best done as a group lab because having each group get the data will take a long time. The roughness of this lab is good, because it shows students that when you work with a real bicycle instead of one which "has all the wheel's mass at the rim," the results get messy. You might try to get data when a student stands on the pedals, but your data will probably be worse still. The reasons for this are that the wheel will turn very fast and the number of revolutions will be hard to measure. Also, it will be difficult for the student to put all his or her weight on the pedal and not have some on the handlebars.

The experiment is rough because lots of energy is lost. The pedals have some kinetic energy and the bearings lose some to friction. One thing which compensates for this error is that all the mass of the wheel is assumed to be at the rim, which is not the case. If you have a wheel which you can take apart, or if you go to a bicycle shop that has a balance, you can figure out the moment of inertia of the wheel more precisely. However, with a more precise value for the moment of inertia (which will be smaller), you will actually find that you have less energy in the wheel.

The support for the bicycle must be rigid. Probably the best support is a device which holds a bicycle stationary for indoor exercise. However, it is possible to suspend the bicycle by rope to the ceiling and then clamp the front wheel to a table leg with a C-clamp. The problem with this setup is that it is difficult for a student to get on the bicycle unless the ceiling support is strong enough and someone holds the front of the bicycle stable.

Friction may cause a noticeable change in velocity in 10 seconds. Test the bicycle to make sure that the bearings are turning smoothly. The friction of the bearings could allow you to extend the lab and find the angular acceleration of the wheel and the frictional torque.

(continued)

Teacher Notes (continued)

Possible Problems

1. Friction is the main source of error and if this is too large, it can be a problem.
2. Getting data from a student on the bicycle can be dangerous.
3. The wheel will slow down during the 10 seconds. Try to make sure that the bearings are turning smoothly.

Conclusions

1. Generally there is less rotational kinetic energy than potential energy.
2. Lost energy to friction and measurement errors are the main sources of error.
3. Energy could go to friction and to rotational kinetic energy lost to the pedals and other moving parts.
4. The energy lost will vary depending on the particular bicycle.
5. The velocity should be higher by a factor of square root of 2 (1.41).

LAB QUIZ ANSWERS

1. a
2. d
3. b
4. c
5. b

Lab #23
Electroscopes

Name: ______________________

Date: ______________________

In this lab you will investigate electroscopes, instruments which allow you to see the effect of electric charge. You will experiment with a vaned electroscope and pith balls which move with small forces because they are very lightweight (see Figure 23). You will experiment with positive and negative electrical charge and learn how to put charge on an electroscope by conduction and induction.

Figure 23

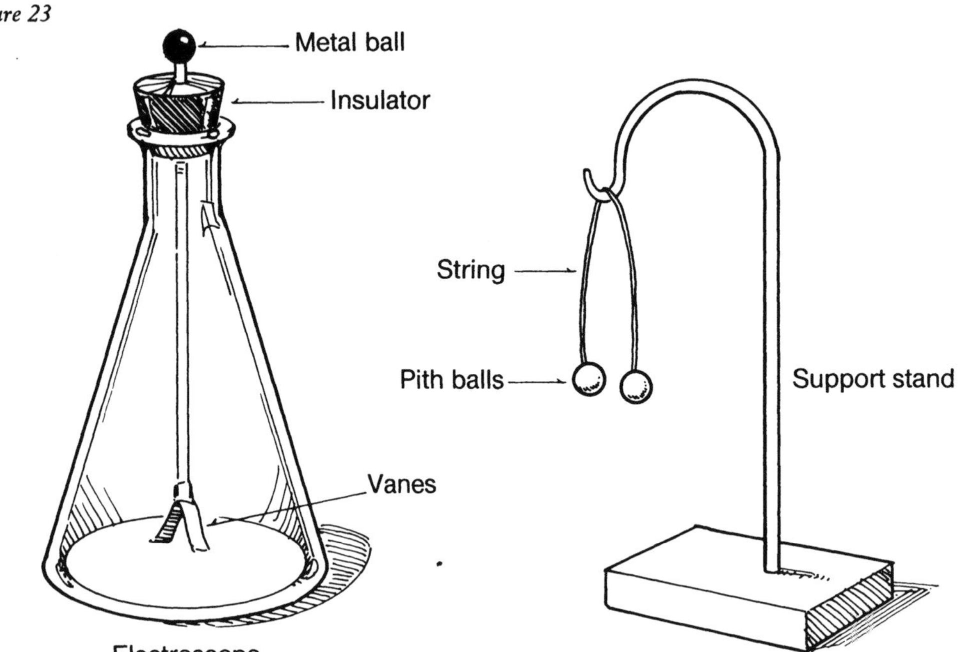

Remember: Like charges repel and unlike charges attract.
Rubber rods rubbed with fur become negatively charged.
Glass rods rubbed with silk become positively charged.

Materials

- rubber rod
- fur
- vaned electroscope
- glass rod
- silk
- pith balls on string with support

(continued)

Lab #23
Electroscopes *(continued)*

Name: ______________________

Date: ______________________

Procedure

1. Rub the rubber rod with the fur. Then touch the rod to the top of the electroscope and rub it back and forth. Record your observations. Why do the leaves (vanes) move as they do?
2. Touch the electroscope with your finger. Record your observations. What must have happened to the charges when you touched the electroscope?
3. Rub the glass rod with the silk. Then touch the rod to the top of the electroscope and rub it back and forth. Record your observations. Why do the leaves move as they do?
4. Touch the electroscope with your finger again. Then bring a charged rubber rod near the electroscope but do not touch it. If your hear a slight snap or the leaves suddenly jump apart, a spark has jumped to the electroscope. This is not desired, so touch the electroscope again and then bring the rod not quite as close to the electroscope as last time. Record your observations. Why do the leaves move as they do?
5. Move the rubber rod away from the electroscope. Record your observations. Why do the leaves move as they do?
6. Now touch the electroscope with a charged rubber rod. The leaves should diverge because negative charge was conducted onto the electroscope. Then rub the rubber rod again and bring it near the top of the electroscope but not so close that a spark jumps. Record your observations. Why do the leaves move as they do?
7. Now rub the glass rod with silk and bring it near the top of the electroscope. Again, do not let a spark jump. Record your observations. What does this mean about the charge on the glass rod as compared to the rubber rod?
8. Discharge the electroscope by touching it with your finger. Then bring a rubber rod near the top of the electroscope but do not allow a spark to jump. With the rod near the top of the electroscope, touch the electroscope with your finger. Do not move the rod away at this time. Record your observations. Why do the leaves move as they do?
9. Now remove the rod from near the top of the electroscope. Record your observations. Why do the leaves move as they do?
10. The leaves should be diverged at this time. Now find out what charge is on the leaves using the results of steps 6 and 7. Bring a charged rubber rod near the top of the electroscope. (Remember the rod is negatively charged.) Record your observations. What charge is on the electroscope? Why?
11. Repeat steps 8–10 but use a charged glass rod instead of a charged rubber rod.

(continued)

12. Now touch the pith balls with the negative rod. They should jump away from the rod. Why do they do this?
13. When the pith balls are apart, bring the section of fur where you have been rubbing the rod near the pith balls. Record your observations. Why do the pith balls move as they do?
14. Roll the pith balls around in your fingers so that when you let them go, the pith balls hang together. What must have happened to the charge on the pith balls?
15. Now slowly bring the charged rubber rod near the pith balls. Record your observations. Why do the pith balls move as they do? What must be happening inside the pith balls?
16. Touch the pith balls with a negatively charged rod. Then find at least two other things which will attract and two things which will repel the pith balls. List your objects and how you charged them.

Caution

Do not hold the silk or fur in your hand for long periods of time. They will become covered with a thin layer of sweat and will not give the rods a large amount of charge.

Conclusions

1. When you touch the top of the electroscope with a charged rod, conduction occurs. How must the charges move?
2. How can you find if the charge on an electroscope is positive or negative?
3. Induction means that charges are forced to move at a distance without conduction occurring. They are induced to move. In which of the situations with which you have experimented does induction occur?
4. You have a charged object. How would you go about finding whether the charge was positive or negative? (Assume that the charge is not large enough to hurt you.)
5. When you touch the electroscope with your hand, you are said to "ground" the electroscope. What must be happening? Why is it not possible for you to "ground" the wires in your house by touching them without danger of being electrocuted?

Quiz #23
Electroscopes

Name: ______________________________

Date: ______________________________

1. If you have two negatively charged pith balls, they will
   a. attract.
   b. attract and then repel.
   c. repel.
   d. repel and then attract.

2. Electroscopes allow you to
   a. exert forces on charges.
   b. magnify charges so they can be seen.
   c. see the effects of electrical charge.
   d. study electric charges under a microscope.

3. A rubber rod rubbed with fur will
   a. pick up negative charge.
   b. pick up positive charge.
   c. pick up negative charge and then become positive.
   d. pick up positive charge and then become negative.

4. In most of the steps in the procedure you are asked to
   a. make distance measurements.
   b. make measurements of the amount of charge.
   c. make measurements of the mass of the electroscope or pith balls.
   d. record your observations.

5. During the experiment, you should avoid
   a. the dangerous shocks which you can get from the electroscope.
   b. having the heavy pith balls fall on you.
   c. getting sweat from your hands on the silk or fur.
   d. none of the above.

Lab #23: Electroscopes

Teacher Notes

Caution

Do not hold the silk or fur in your hand for long periods of time. They will become covered with a thin layer of sweat and will not give the rods a large amount of charge.

In this lab students investigate conduction and induction and see how the electroscope can be used to identify charges. Students are led through the steps, which you may already have done as a demonstration in class. Sometimes, however, having students do the steps in the lab is much more beneficial.

The equipment is standard and a complete lab set can be purchased for about $25 from most physics supply houses. Because pith balls can be bought and mounted on a variety of supports, the supply house support is not really needed (although it may make the lab materials look better).

Possible Problems

1. If too much sweat from the student's hands gets on the fur or silk, they will not give a good charge.
2. If the day is humid, it is hard to charge the electroscopes.
3. If there is any connection between the electroscope top and leaves to the electroscope case, the electroscope will not become charged.
4. Electricity is fickle and sometimes it just does not work as planned. This lab can then become frustrating for the teacher. However, if problems 1 and 2 are avoided, things generally work well.

Conclusions

1. Charges move directly from one place to another. They move on or off the electroscope because of electrical forces.
2. Bring a known charge near the electroscope. If the leaves diverge more than they were, then the electroscope has the same charge. If the leaves move together, then the charge on the electroscope is the opposite charge.
3. Induction occurs whenever you bring the charged rod near but do not touch the pith ball or electroscope.

(continued)

Teacher Notes (continued)

4. The best way is to charge some pith balls with a known charge and then bring them near the charged object. If they move toward the unknown object, it has opposite charge. If they move away, the unknown mass has the same charge.
5. The electrical system in the house can give you much more current than any object in this electrostatic experiment. If too much current passes through your body, you risk injury or death.

LAB QUIZ ANSWERS

1. c
2. c
3. a
4. d
5. c

Lab #24
Balloons and Coulomb's Law

Name: ______________________

Date: ______________________

In this lab you will investigate the charge on two balloons. When balloons are rubbed on hair, they pick up charge. This property of rubber balloons is used at many birthday parties, but how much charge is on the balloon? By using two balloons and making some simple measurements, it is possible to find the charge on the balloons.

Materials	
balloons	protractor
string	meter stick
support	

Procedure

1. Blow up and tie the ends of two balloons. Tie a string to each of them and then tie the other end of the strings to the support so that both strings come from a point. The balloons should be the same distance below the support.
2. Measure the distance from the support to the center of a balloon. (This should be the same distance for the other balloon.)
3. Now rub each balloon on your hair until it picks up charge. With charge, they should separate just like pith balls with similar charge (see Figure 24.1). Rub the balloons again and try to get the largest separation possible. This means that you have the largest charge on each balloon possible.
4. With a protractor, measure the angle between the strings of the balloons and measure the distance between the centers of the balloons. Record these values. With the balloons charged, it is a bit difficult to measure distances and angles because the balloons may move toward the measuring device. If you have plastic rulers and protractors, try rubbing them on your hair to give them negative charge. This may help you to get better measurements.
5. Find the mass of the balloons using a sensitive balance or use the value for mass given by your teacher.

Analysis

1. Figure 24.2 can be used to derive the following equation. The balloon must move tangent to the circle at radius l, the length from the support to the center of the

(continued)

Name: ______________________________

Date: ______________________________

Figure 24.1

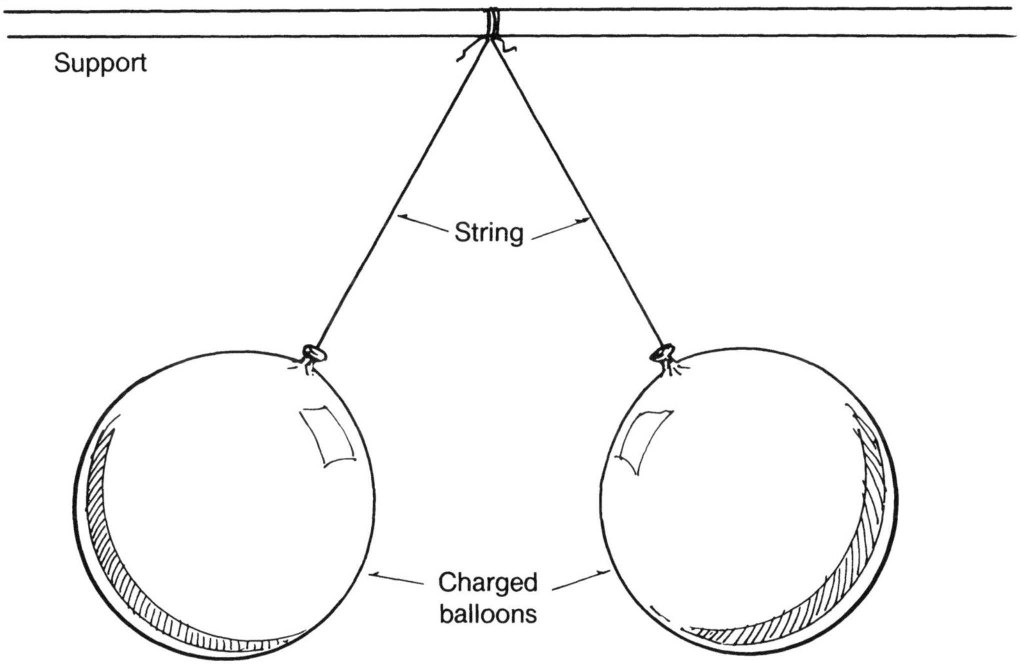

balloon. Therefore, only the components of the electrical and gravitational force tangent to the circle are important. These components are set equal to each other in the equation:

$$\frac{kq^2}{r^2}\cos\left(\frac{\theta}{2}\right) = mg\sin\left(\frac{\theta}{2}\right)$$

where m is the mass of the one balloon, g is the acceleration due to gravity, θ is the angle between the balloons, k is the electrical constant, q is the charge on a balloon, and r is the separation between the balloons. You measured m, θ, and r; g equals 10 m/sec^2 and $k = 9 \times 10^9 \frac{n \cdot m^2}{C^2}$. The only unknown in the equation is q, the charge on the balloon. Make sure your measurements are in meters and kilograms and then solve for this charge.

2. Given that there are 6.2×10^{18} electrons/coulomb, find the number of electrons on each balloon.

(continued)

Lab #24
Balloons and Coulomb's Law
(continued)

Name: ______________________

Date: ______________________

Figure 24.2

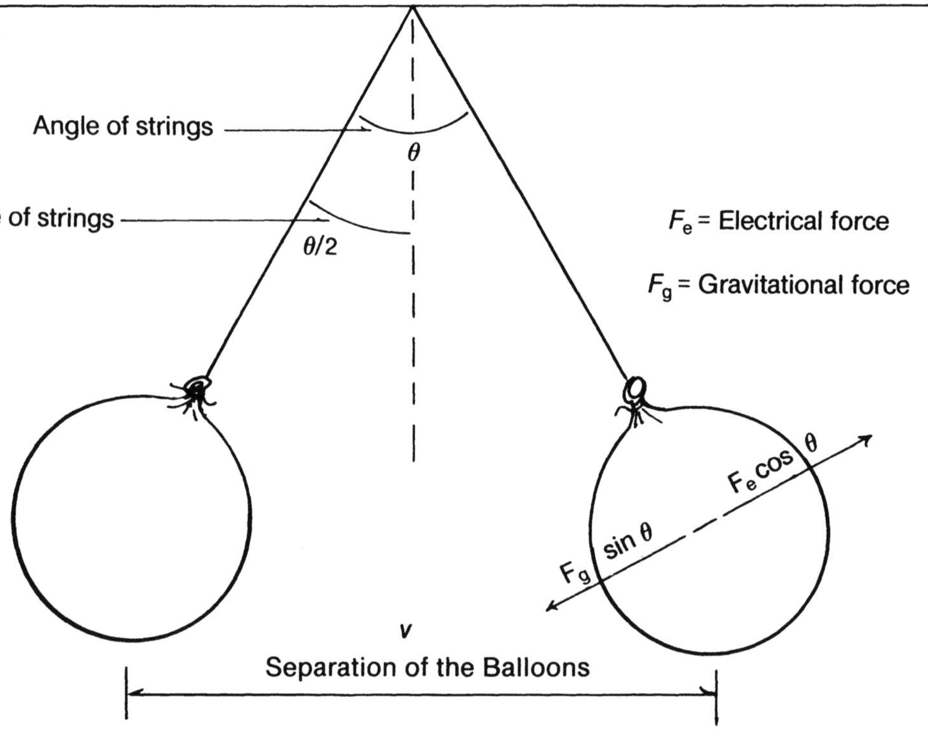

3. Give a rough estimate of how much error there is in each of your measurements and use these values to estimate the error in the charge.

Conclusions

1. Does the amount of charge on the balloon surprise you? Explain.
2. Explain how the equation used to find the charge is derived.

Quiz #24
Balloons and Coulomb's Law

Name: ______________________________

Date: ______________________________

1. Charge is placed on the balloons by

   a. rubbing them on hair.
   b. touching them with charged rubber rods.
   c. touching them with charged glass rods.
   d. rubbing them against a highly charged object.

2. In some ways this experiment is similar to

   a. the experiment you have done at birthday parties with balloons.
   b. the experiment on centripetal force.
   c. the experiment on gravitational acceleration.
   d. the experiment of torque.

3. In this lab a component of the repulsive electrical force between the balloons is balanced by a component of the

   a. spring force.
   b. frictional force.
   c. gravitational force.
   d. negative electrical force.

4. A component of a force is a ____________________ of a force.

   a. section
   b. part
   c. unit
   d. definition

5. You rub the balloons trying to get the largest separation possible because that will be associated with the largest ____________________ possible.

   a. mass
   b. charge
   c. length of string
   d. time

Lab #24: Balloons and Coulomb's Law

Teacher Notes

In this lab students calculate the charge on a balloon. The charged balloon is the same type students have rubbed on their heads and stuck on the walls at birthday parties. For this reason, it is interesting to find the large number of electrons actually transferred from hair to balloon and it is good physics, too.

The lab gives a rough idea of the total charge on the balloon. There are large measurement errors mostly resulting from the motion of the balloons when approached with a protractor or ruler. Also, the charge distribution is not symmetric. If anything, however, the charge is larger because the balloons should turn so that the side with the smallest charge is toward the other balloon.

The balloons should be round ones so that the distribution of charge is relatively spherical. The mass of the balloons can be measured by students, but it is best done by finding the average mass of a number of balloons because the mass of a single balloon is small. If you have access to a balance which measures hundredths of grams, then you can find the individual masses of the balloons.

Possible Problems

1. Students may become confused with the equation. Perhaps just have the weaker math students put in the proper values for the unknowns. You may want to do the derivation slowly on the board.
2. The measurements are difficult to make accurately because of the motion of the balloons when the measurement devices are brought near. Rubbing the devices on hair does sometimes help in getting better measurements. Encourage the students to look for ways to get as precise measurements as possible.

Conclusions

1. The number of coulombs is small but the number of electrons is huge. This surprises some students.
2. The derivation is discussed in the analysis section but it will confuse many students. Nevertheless, they should be able to state that the component of the electrical force which moves the balloons along the arc ($f \cos\theta$) equals the component of gravity ($f \sin\theta$) which tends to move them together.

(continued)

Teacher Notes (continued)

Lab Quiz Answers

1. a
2. a
3. c
4. b
5. b

Lab #25
Simple Circuits

Name: ______________________

Date: ______________________

In this lab you will investigate current in parallel and series circuits. You will use a power source, resistors, ammeters, and voltmeters to see if the current and voltage which are predicted by equations actually exist in circuits. You will measure the voltage drop across the resistors and the current through them in three circuits. You will compare the predicted values to the measured values.

Materials

wires	three resistors
voltmeter	ammeter
low-voltage power supply	

Procedure

Note: Before you plug your circuit into the power supply, make sure that you have constructed it properly. Check with your teacher if you have questions.

1. Plug in the two wires to your voltmeter and then plug the other ends into the power supply (see Figure 25.1). Make sure the negative side of the voltmeter is plugged into the negative side of the power supply. Turn on the power supply and adjust the voltage to the level indicated by your teacher. Remove the voltmeter and leave the power supply on.
2. Using two resistors, make a series circuit as shown in Figure 25.2. Attach two wires which will eventually be plugged into the power supply.
3. Place wires from the voltmeter *across* one of the resistors. One wire from the voltmeter is placed on one side of the resistor and the other is placed on the other. Make sure that the negative side of the voltmeter will be closer to the negative side of the power supply than the positive side. Otherwise the meter will move the wrong way. Do not plug the circuit in until you know you have the circuit constructed properly.
4. When your circuit is properly constructed, plug it into the power supply and read the voltage drop across the resistor. Record this value and unplug your circuit.
5. Move the voltmeter to the other resistor and repeat step 4 to find the voltage drop across the other resistor. Record this value. Do the values make sense, given that the resistors are the same size?

(continued)

Name: ______________________________

Date: ______________________________

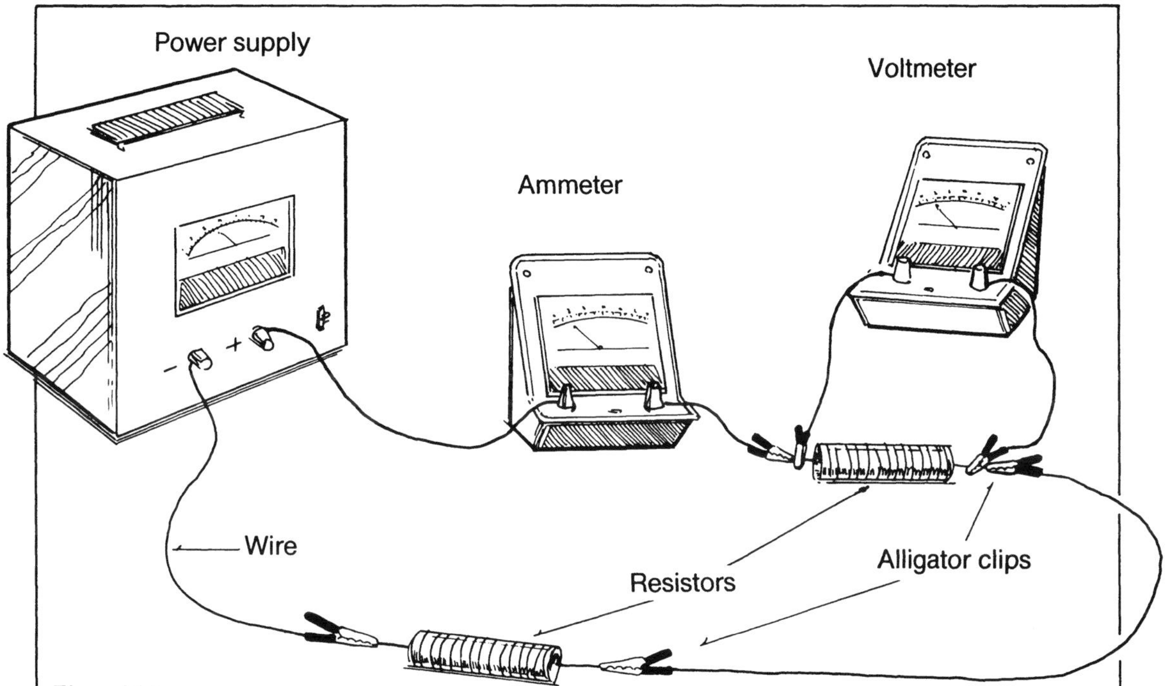

Figure 25.1

6. With the circuit unplugged from the power supply, *break* the circuit right next to one of the resistors by taking the clip off the wire that is at one end of the resistor. Use the two wires coming from the ammeter to connect the circuit again. Remember to have the negative side of the meter near the negative side of the power supply.
7. When you have connected the circuit properly, plug it into the power supply and read the current flowing at the point where you broke the circuit. Record this value and unplug the circuit from the power supply.
8. Repeat steps 6 and 7 but this time place the ammeter next to the other resistor.
9. Now construct a circuit with two resistors in parallel as shown in Figure 25.2b. Using the procedure outlined in steps 3–8, find the voltage drop across and the current through each resistor. Make sure that you find the current through the entire circuit *and* the current through each parallel resistor. Remember to *break* the circuit when you place the ammeter in the circuit.
10. Finally, construct a circuit with two resistors in parallel and one in series as shown in Figure 25.2c. Using the procedure outlined in steps 3–8, find the voltage drop across and the current through each resistor. Make sure that you

(continued)

Name: ______________________________

Date: ______________________________

Figure 25.2

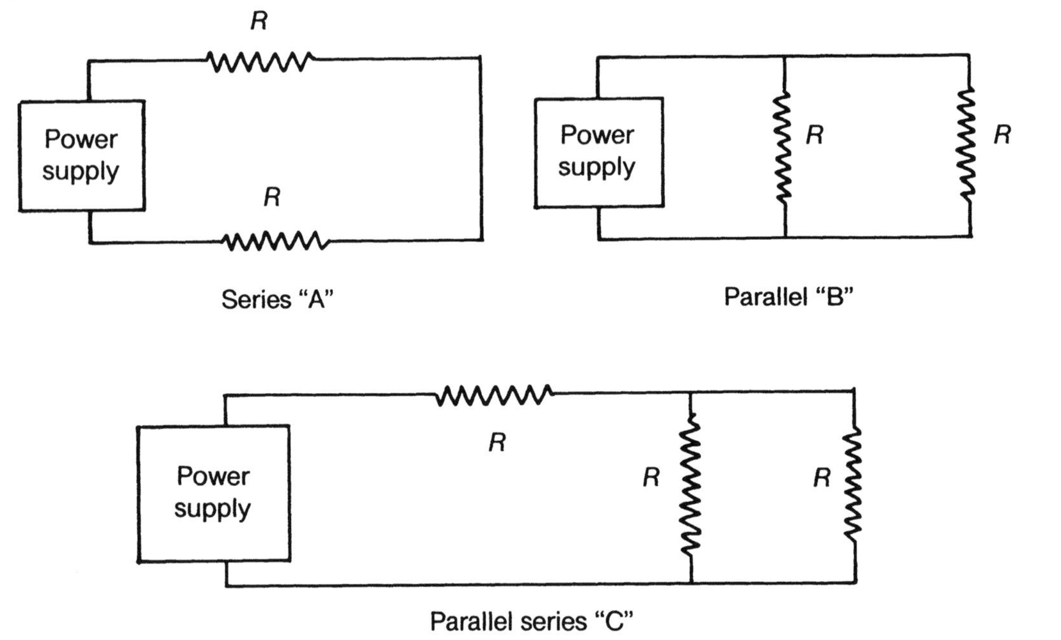

find the current through the entire circuit *and* the current through each parallel resistor.

Cautions

1. To avoid burning out a fuse in your ammeter, make sure that the ammeter is always connected properly before plugging the circuit into the power supply. You may not have thought you did anything wrong or it may have only been for "a second," but that is enough to blow a fuse.
2. Make sure that the negative side of the meter is always closest to the negative side of the power supply.

(continued)

Lab #25

Simple Circuits *(continued)*

Name: ______________________________

Date: ______________________________

Analysis

Calculate the amount of current which should flow through and the voltage drop across each resistor in each circuit.

Conclusions

Compare the calculated values for the voltage drop and current to the actual values. Do they compare? Are they reasonable?

Quiz #25
Simple Circuits

Name: ______________________

Date: ______________________

1. A voltmeter measures ____________ a resistor.
   a. current across
   b. current through
   c. voltage across
   d. voltage through

2. An ammeter measures ____________ a resistor.
   a. current across
   b. current through
   c. voltage across
   d. voltage through

3. When an ammeter is connected in the circuit, you always ____________ the circuit.
   a. break
   b. fix
   c. apply
   d. plug in

4. If there is a voltage drop of 6 volts across a 2-ohm resistor, the current flowing in it is
   a. 1 amp.
   b. 2 amps.
   c. 3 amps.
   d. 6 amps.

5. Two 10-ohm resistors are placed in parallel and then are attached to another 10-ohm resistor in series as shown. What is the net resistance?

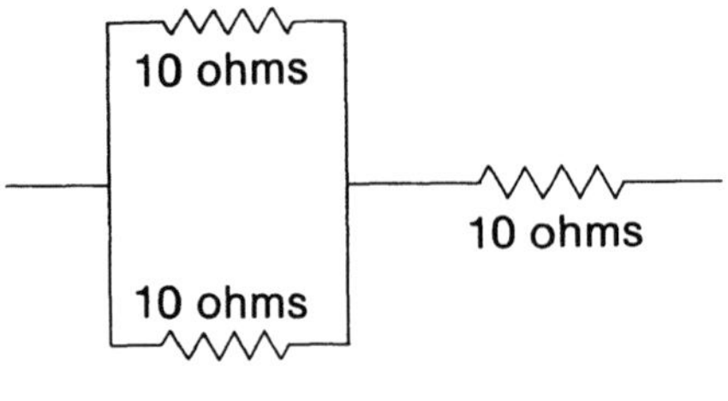

   a. 5 ohms
   b. 10 ohms
   c. 15 ohms
   d. 30 ohms

Lab #25: Simple Circuits

Teacher Notes

Cautions

1. To avoid burning out a fuse in your ammeter, make sure that the ammeter is always connected properly before plugging the circuit into the power supply. You may not have thought you did anything wrong or it may have only been for "a second," but that is enough to blow a fuse.
2. Make sure that the negative side of the meter is always closest to the negative side of the power supply.

In this lab students look at current and voltage in simple circuits and learn that the equations work. The only errors result from the fact that the voltage from some power supplies may vary when a load is placed across them and that the needles on the meters may not be zeroed. Because the students should know something about error by this time, they should have no problem with these errors and should see that the lab shows the equations to be correct.

Before the lab starts, explain how to make the connections. The resistors have a wire at each end which can be used for making connections. Show the students how to twist them together and how to put them into the patterns shown in each of the three circuits. Tell them that a voltmeter measures voltage *across* a resistor and therefore should be placed *across* a resistor. Tell them that an ammeter measures current through a point in the circuit. Since it is placed into the circuit, the students must *break* the circuit and connect the break with the ammeter. *Stress* the word *break*. This will help when students are not doing the steps correctly and have burned out the fuse on their ammeter. Also tell students that the negative side of the meters must be closer to the negative side of the power supply than the positive side, and explain how to identify which is which.

The equipment for this lab must be of reasonably good quality or you will have difficulty giving students the help they will need building the circuits.

The following items are needed for the lab:

1. A good voltage source which can easily provide 1 amp of current without overloading and which has a stable voltage for a range of currents. It should also be fused so that it does not overload. It may be possible to have two or more groups use the same supply if the supply is properly protected and maintains a constant voltage. Variable voltage to 12 volts and 3 amps is desirable (although 6 volts is sufficient). Such power supplies are available from supply houses.

(continued)

Teacher Notes (continued)

2. Voltmeters which read to a larger voltage than the maximum output from your power supply. Generally, 0–15 volts is fine.
3. A fused ammeter or one with a circuit breaker which reads to 1 amp. Without the fuse or breaker, you will destroy meters. If they have fuses, make sure that you have extra fuses to replace the burned-out ones.
4. Resistors in the range of 10 ohms to 50 ohms. It is easiest if they are all the same size. They should be able to handle 2 watts of power. You are aiming for .2 to .4 amp though a resistor if it is placed directly across the power supply. This means that you must adjust voltage to the resistor size. If you use 10-ohm resistors, use about 4 volts. If you use 50-ohm resistors, use about 10 volts. With more voltage, the resistors get too hot for the students to handle; or with less voltage, the readings on the meters are so low that students don't believe them.

Possible Problems

1. Students may blow fuses.
2. Students may be confused as to how to set up their circuits.
3. Poor equipment may cause you to spend more time repairing the equipment than helping students.

Conclusions

The currents and voltage drops should compare well unless the power supply changes its voltage with a load or the meters are not sensitive enough.

LAB QUIZ ANSWERS

1. c
2. b
3. a
4. c
5. c

Lab #26
Drawing Magnetic Fields

Name: ______________________

Date: ______________________

In this lab you will investigate magnetic fields around magnets and wires. You will use a small compass to find the direction of the magnetic field and you will plot this field on a piece of paper.

Materials	
magnets	power supply
paper	wire
pencil	coil of wire
small compass	

Procedure

Case 1:

1. Place a bar magnet on a piece of paper and draw an outline of it with a pencil (see Figure 26.1). Then place the small compass near the magnet and note the direction of the north-pointing arrow. Remove the compass and draw an arrow pointing in the same direction.
2. Repeat this process until you have a diagram showing the direction of the field within 10 cm of the magnet. Remember that if the field from the magnet is weak, the needle will point toward the magnetic North Pole. Also, if other magnets are nearby, they will affect the direction in which the compass points.
3. Remove the magnet and then place the compass in a corner of your paper. The direction in which the compass points is magnetic north if no magnet is near. Draw an arrow in this direction and label it *north.*

Case 2:

4. Place a horseshoe magnet on a piece of paper and draw an outline of it (see Figure 26.1). Follow the instructions in steps 1, 2, and 3 so that you can draw a diagram of the field around the magnet.

Case 3:

5. You will now find the magnetic field around a wire which is perpendicular to a plane. Put the wire across the center of the paper (see Figure 26.2). With the power off, place your compass on the board. Find out which way is north from

(continued)

Name: ______________________

Date: ______________________

Figure 26.1

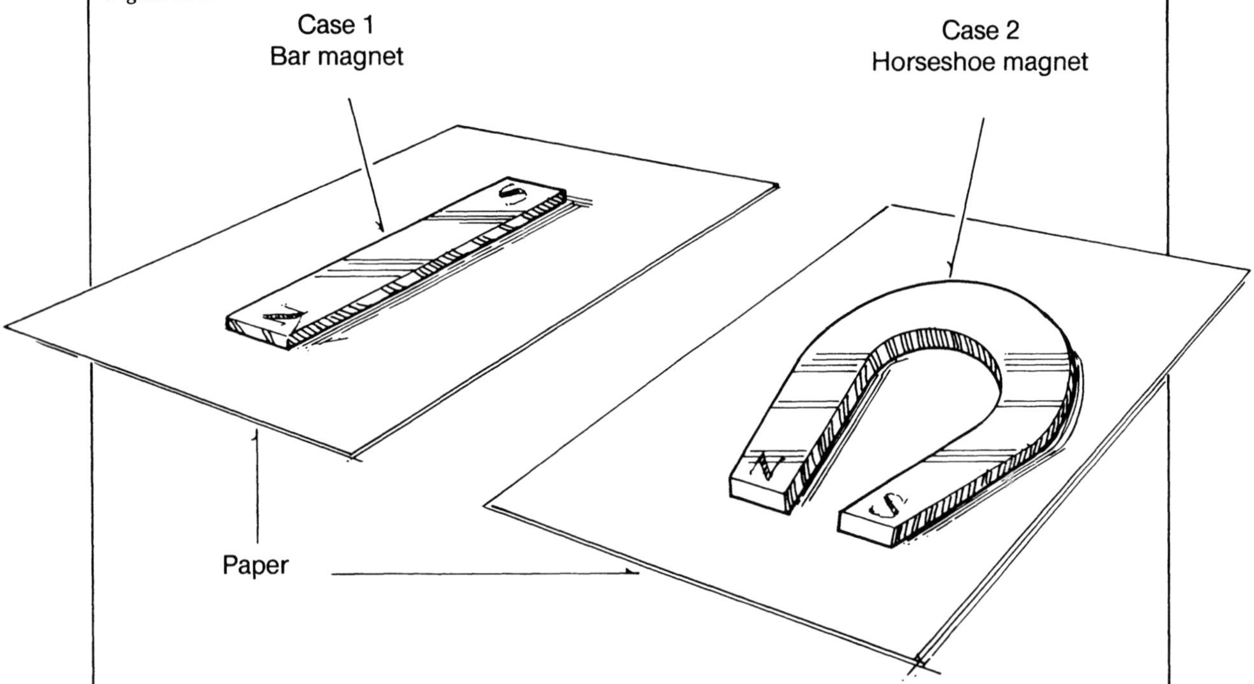

the compass. Move your compass so that it is north of the vertical wire and about 1 cm from the wire.

6. Turn on the power supply and slowly increase the current to the maximum value given by your teacher. Observe what happens to the needle and record your observations. Then move the compass around and plot the magnetic field by following the procedure outlined in steps 1, 2, and 3. When you have completed your diagram, turn off the power supply.

Case 4:

7. With the power off, place a piece of paper under the coil of wire (see Figure 26.3). Draw an outline of the coil and then turn on the power supply to the value given by your teacher. Draw a diagram of the field using the procedure outlined in steps 1, 2, and 3.

(continued)

Name: ______________________

Date: ______________________

Figure 26.2

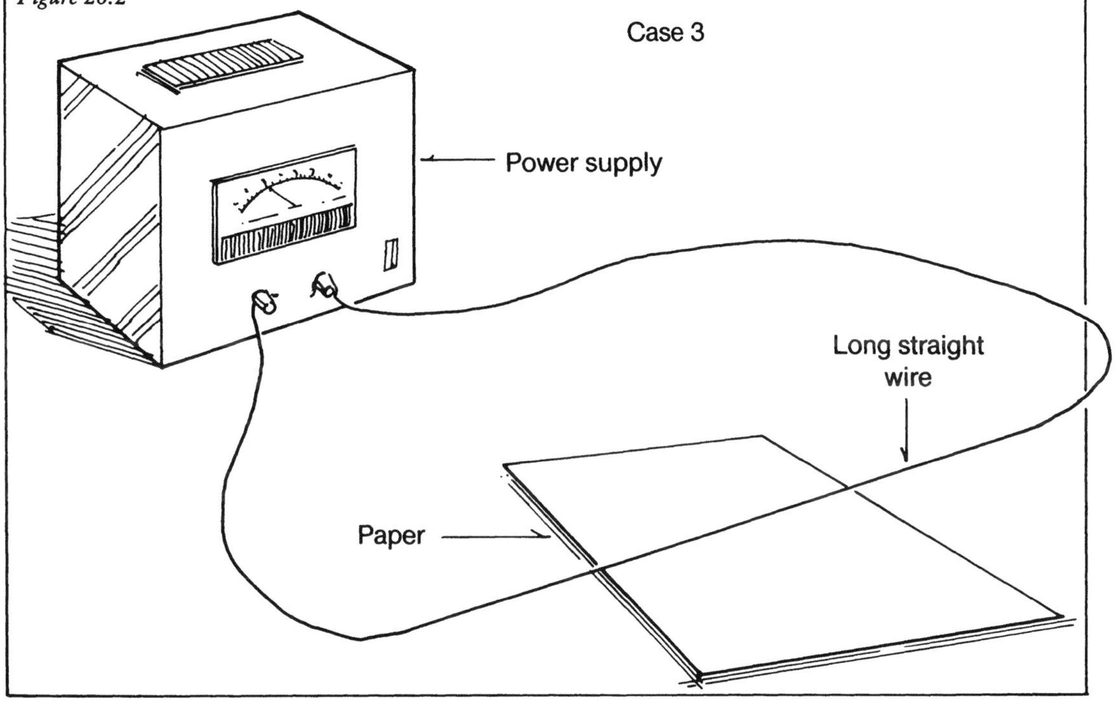

Caution

Do not use more current than your teacher gives as the maximum value. You may blow out a fuse which will need to be replaced.

Conclusions

1. How do your diagrams compare with the magnetic fields which you expected? Explain.
2. How does the earth's magnetic field affect the measurements which you made? What effect does it have on the diagrams? Explain.
3. Is there any way you could get rid of the effect of the earth's magnetic field? Explain.

(continued)

Name: ______________________

Date: ______________________

Figure 26.3

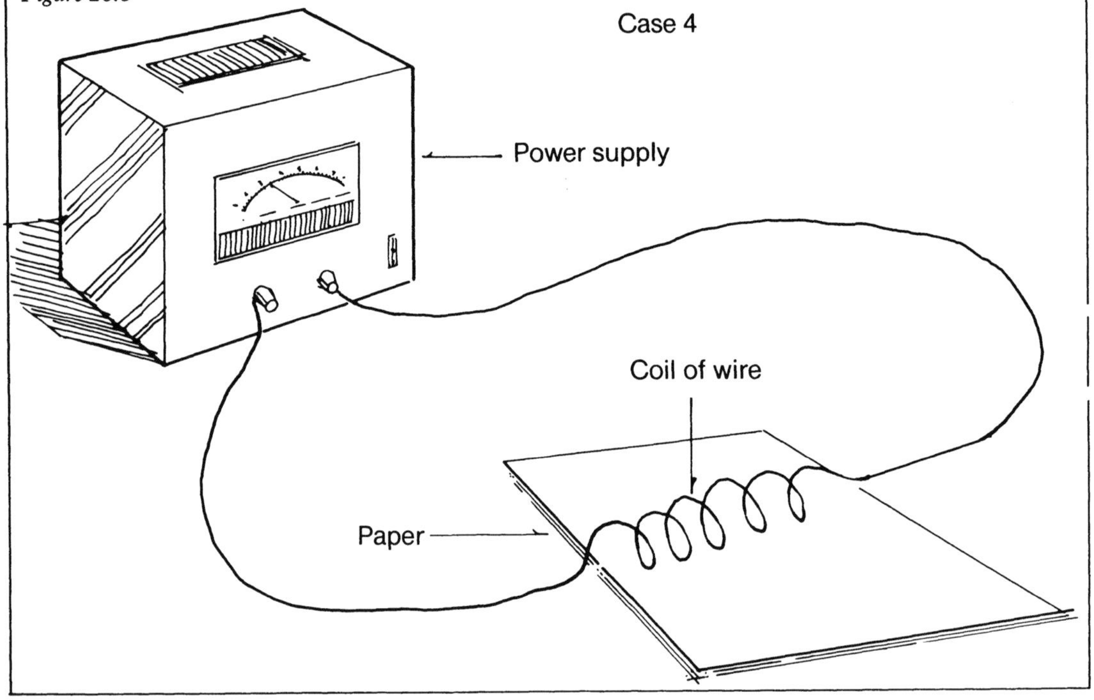

Quiz #26
Drawing Magnetic Fields

Name: ______________________

Date: ______________________

1. A small compass needle in the earth's magnetic field points
   a. north.
   b. south.
   c. east.
   d. west.

2. The direction which the compass needle points will be from the combination of the magnetic field from the magnet, coil, or wire and
   a. any other magnet in the area.
   b. the earth's magnetic field.
   c. both the earth's field and the field of other magnets.
   d. no other magnetic fields.

3. The small compass will be used to
   a. find the direction of the electric field.
   b. find the direction of the magnetic field.
   c. find the strength of the electric field.
   d. find the strength of the magnetic field.

4. If you use too much current in your straight wire or the coil, you could damage the power supply or
   a. ruin your compass.
   b. decrease the magnetic field to zero.
   c. make the magnetic field go the wrong way.
   d. blow a fuse.

5. You place the compass at many places around the magnet, wire, or coil so that you can do all *except* which one of the following?
   a. Draw a complete diagram of the field.
   b. Get more numbers to enter into the equation.
   c. Get a better "feel" for the magnetic field.
   d. Find if there are places where the field is not as you would have expected.

Lab #26: Drawing Magnetic Fields

Teacher Notes

Caution

Do not use more current than your teacher gives as the maximum value. You may blow out a fuse which will need to be replaced.

This lab helps students to visualize magnetic fields and to see how the fields can be affected by the earth's field. With the proper equipment, it is possible to do this lab quickly with good results. It is not a numerical lab and sometimes students like to have these as a change from ones with lots of calculations and data collecting.

There is no need to have the students plot the magnetic fields in any particular order. If you do not have enough setups for each student, you can start some on magnets, some on straight wires, and some on coils of wire. They can then move to the next station.

The equipment is simple except for the power supplies and the setup of the vertical wire. Small 1-cm-diameter compasses work best for this lab. Regular bar and horseshoe magnets are also appropriate. The power supply for the vertical wire is a problem for some labs because the current required to produce a field large enough to be measured is sometimes not available. Currents of 10 amps or more are required. The wire must be able to handle this current. (Such apparatus is available from physics supply houses.) The maximum value which the students can use for current should be stated before the lab begins and also posted on the power supplies. You can reduce the power requirements dramatically if you use a large square coil of wire which is 20 cm or more on a side. This increases the field and acts just like a single vertical wire. The power supply for the coil does not need to be as big because the coil produces much larger fields. Make sure that you have experimented with the apparatus so you can state the maximum values for voltage and that you are sure students will get good data when finding the field around the straight wire. The straight wire can be supported by a stand and then run down through a hole in the center of a piece of plywood. The plywood is placed over the edge of a table and weighted to prevent it from falling.

Possible Problems

1. If the power supply to the vertical wire cannot make a large enough field, the earth's field becomes dominant and the diagram does not show a circular field. Of

(continued)

course, at great distances from the wire, the compass will point north, but this change from circular to north pointing is important for students to see.

2. The field in some cases becomes zero because of the combination of the earth's field and the other magnetic field. This may confuse students.
3. Students may not take enough points of data. They must look at the field in all regions within 10 cm from the magnet, wire, or coil.

Conclusions

1. The fields should be the expected fields except the earth's field may make some differences.
2. There is always a component of the field in the direction of the earth's field. If the magnetic field created by the magnet, wire, or coil is too small, then the compass needle will point north.
3. It would be difficult, but you could do the experiment near coils of wire with current adjusted to just balance the earth's field. Helmholtz coils provide just such a field.

LAB QUIZ ANSWERS

1. a
2. c
3. b
4. d
5. b

Lab #27
A Current Balance

Name: ____________________

Date: ____________________

In this lab you will investigate forces between two current-carrying wires. A movable wire is placed near a coil and current is run through both the coil and the movable wire. A magnetic force causes a torque on the movable wire which is balanced by an opposite torque provided by a gravitational force. Knowing the gravitational torque which just balances the magnetic torque allows you to calculate the magnetic force.

Materials

movable wire	connecting wires
coil of wire	power supply
meter stick	10-amp ammeter
pencil	masses

Procedure

1. Place the movable wire on top of the apparatus which holds the coil. Adjust the movable mass so that the rod holding the mass is horizontal.
2. Adjust the movable wire so it comes to rest about 1 cm from the center of the coil. Measure this distance precisely and make a mark with a pencil so you can always get the wire back to this same position. Notice that the movable wire is supported on metal plates and even when the wire moves, electricity can still flow.
3. With the power supply off, connect wires so that current will flow from the power supply, through the movable wire, through the coil, through the ammeter, and back to the power supply. Make sure that the negative (black) side of the ammeter is closest to the negative side of the power supply. (See Figure 27 for how to make these connections or ask your teacher for help.)
4. The forces between two wires can be attractive if the current is going in the same direction or repulsive if it is going in opposite directions. You must have a repulsive force for this lab. When your circuit is connected properly, turn on the power and increase the current to about 1 amp. If the movable wire moves away from the coil, then things are working properly. If the movable wire moves toward the coil, then you must reverse the wires attached to the movable wire.

(continued)

Name: ______________________

Date: ______________________

Turn off the power, switch the wires, and turn on the power to check for repulsion. Turn off the power.

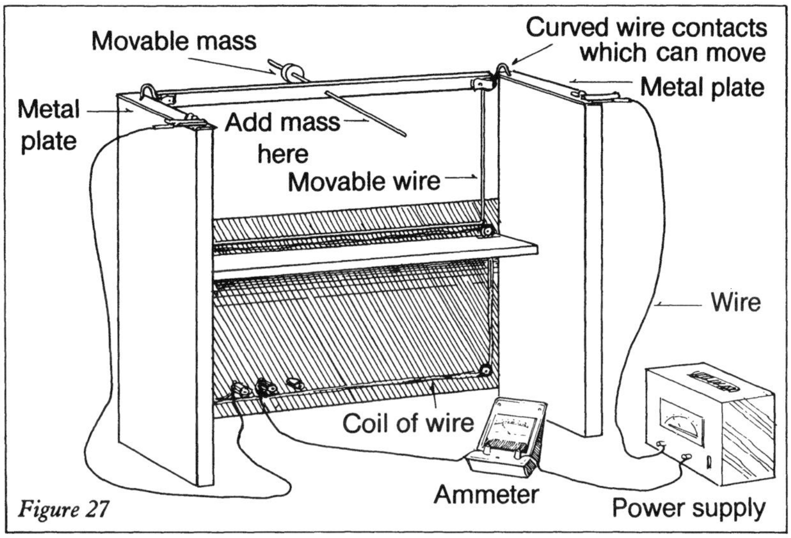

Figure 27

5. Place a mass at the opposite end of the rod from the movable weight. Then turn on the power and slowly increase the current until the wire moves back to its original position (where you made a mark in step 1). Record the amount of mass and the current.
6. Increase the mass as indicated by your teacher, recording the mass and current in each case.
7. Turn off the power supply and disconnect all the wires.
8. Measure and record the following: length of the movable wire which is next to the coil, distance from the pivot point to the movable wire, distance from the pivot point to where you placed your mass, the separation of the wires, and the mass of each weight.
9. (Optional) Put the long movable wire back onto the apparatus. Move the wires connected to the coil so that they attach to just one wire carrying current instead of ten. Then place one mass on the rod at a distance which is 1/10th the distance from the pivot point where the other masses have been placed. Carefully adjust the current so the movable wire returns to its original position.

(continued)

Name: ______________________

Date: ______________________

Caution

Do not use more current than your teacher recommends. More than this amount will blow a fuse or a circuit breaker. The voltage which you are using is not dangerous, but you can harm the power supply if you are not careful.

Analysis

1. Find the gravitational force exerted by one weight.
2. Find the torque exerted by five weights using the gravitational force on five weights and the distance from the pivot to where the weights were placed (the moment arm).
3. The gravitational torque is balanced by a magnetic torque. The magnetic torque is produced by the magnetic force on the movable wire times the distance from the pivot to the movable wire (its moment arm). Set the gravitational torque equal to the magnetic torque and find the magnetic force on the movable wire.
4. The magnetic force between two wires is given as:

$$f = \frac{2kl\,I_1\,I_2}{s}$$

   where f is the force on each wire in N; k is a constant equal to 10^{-7} N/amp^2; l is the length of the wires in m; I_1 and I_2 are the currents in each wire in amps; and s is the separation distance in m.

   Use this equation to find the magnetic force on the movable wire. l is the length of the movable wire. I_1 is the current in the movable wire which is the current on the ammeter. I_2 is the current in the coil of wire which has 10 turns of wire so the current is 10 times the current on the ammeter. s is the separation between the wires. (Make sure that you are using meters and not centimeters for measurements.) Use the current when you used 5 masses.
5. Plot a graph of number of weights vs. current for the values of mass and current which you recorded.

(continued)

Name: ______________________________

Date: ______________________________

Conclusions

1. How well does the magnetic force calculated using torques compare to the magnetic force calculated with the equation? Are the differences reasonable? Explain.
2. When you make twice as much current flow from the power supply, twice as much current goes through the movable wire *and* the coil. If both of these are doubled, the equation indicates that the force should be four times as great. Does your graph show that this is the case? Explain.
3. (Optional) When you make the current only go through one wire in the coil, you reduce the current in the coil by a factor of 10. Did your experiment show this?

Quiz #27
A Current Balance

Name: ______________________

Date: ______________________

1. Current flowing in parallel wires which are separated by a small distance will exert ______________ on each other.

   a. an electric force
   b. a magnetic force
   c. an electric potential
   d. a magnetic potential

2. The apparatus which is used in this lab includes a fixed coil of wire and a

   a. movable wire.
   b. fixed wire.
   c. wire perpendicular to the coil.
   d. wire in the shape of a spring.

3. Torque is the product of

   a. force times mass.
   b. mass times weight.
   c. force times distance.
   d. force times velocity.

4. An object will not change its rotational motion if the

   a. forces are balanced.
   b. weights are balanced.
   c. torques are balanced.
   d. force is fixed.

5. A wire is in a magnetic field. If the current in the wire is doubled, the force on the wire is

   a. halved.
   b. the same.
   c. doubled.
   d. quadrupled.

Lab #27: A Current Balance

Teacher Notes

> ***Caution***
>
> Do not use more current than your teacher recommends. More than this amount will blow a fuse or a circuit breaker. The voltage which you are using is not dangerous, but you can harm the power supply if you are not careful.

This is a good lab for students in that they actually experiment with magnetic forces which they have very little experience with in daily life. Of course, any electric motor works by exerting forces on wires, but it is not easy to see that happening. This lab can be expanded (using the "optional" section) to investigate the effect of changing the amount of current in one of the wires as opposed to changing it in both. The commercial apparatus supplies materials to investigate this effect.

The apparatus is available from physics supply houses for about $50 per setup. It is difficult to make your own, but it is possible. The power supply for the lab must produce at least 5 amps because the forces between the wires are so small. Larger currents are helpful.

The masses used for this lab should be made specifically for the lab. They should be small segments of thin wire (or even pieces of paper) with a very small mass. The best way to select their mass is to set up the lab and find the maximum amount of current which your power supply can handle attached to the apparatus. Then find how much mass needs to be added to the end of the horizontal wire in order to balance the magnetic force by adding wire or paper until it balances. When you find the total amount, make your standard of mass equal to 1/5th of this total. Cut other pieces of wire or paper so that you have a supply of these masses. Find the mass by putting the total mass on a sensitive balance and divide by the total number. It is possible to use the movable mass which comes with the equipment. However, as students adjust it, they often move the wire from its original position and error is introduced. Adding and removing weights from the horizontal wire is much easier and more precise. It is also possible to glue a small mirror to the movable wire and shine a light on it. Students then adjust the current so the light reflected from the mirror is in its original position.

Part of this lab has been made optional to avoid overloading students. Use judgment to decide if they can handle it. It is also possible to change the length of the movable wire. The standard apparatus provides such a wire.

(continued)

Teacher Notes (continued)

Possible Problems

1. If the movable wire is bumped, the pivot point will change and the data will be poor even if the students put it back where it was.

2. The equation is quite threatening and the calculations may mystify some students. The combination of a new equation and having to use torques makes this lab difficult for some students.

Conclusions

1. The torques should be equal although there may be a 20% error due to not getting the movable wire back to its original position. This is hard to do.
2. Yes, with double the current, four times the force should occur. It is a parabolic curve.
3. The error in this lab is generally large because of the small numbers involved.

LAB QUIZ ANSWERS

1. b
2. a
3. c
4. c
5. c

Lab #28
Electrical Equivalent of Heat

Name: ______________________

Date: ______________________

In this lab you will investigate how electrical energy is changed into heat energy. Everyone knows that electricity can be used to heat things, but the exact relationship is not known to most people. In this lab a small amount of water will be heated with electricity. The heat gained by the water should equal the electrical energy given by the power source.

Materials

balance	voltmeter
calorimeter	heating coil
water	wires
power supply	thermometer
wires	timer
ammeter	

Procedure

1. Find and record the mass of the empty calorimeter (see Figure 28). Then add cold water until the calorimeter is about two-thirds full. Find the new mass and record it.
2. With the power supply off, use wires to make a circuit which goes from the power supply through the ammeter, through the heating coil, and back to the power supply (see Figure 28). Make sure that the negative (black) side of the ammeter is nearest the negative side of the power supply.
3. Connect the voltmeter so the voltage across the heating coil can be measured. Make sure that the negative side of the voltmeter is nearest the negative side of the power supply. *Make sure the voltmeter, ammeter, and power supply do not get wet. They may be damaged or can give electrical shocks.*
4. Place the heating coil in the water, making sure that the calorimeter is stable so that it won't tip over. Then turn on the power supply and adjust the dial so that you get approximately the values which your teacher suggested. Turn off the power supply. When you turn on the power supply again, you will want the same values for voltage and current when you gather your data. *Do not ever turn the power on without having the coil in the water. It will get hot enough to burn you very quickly.*

(continued)

Name: ____________________

Date: ____________________

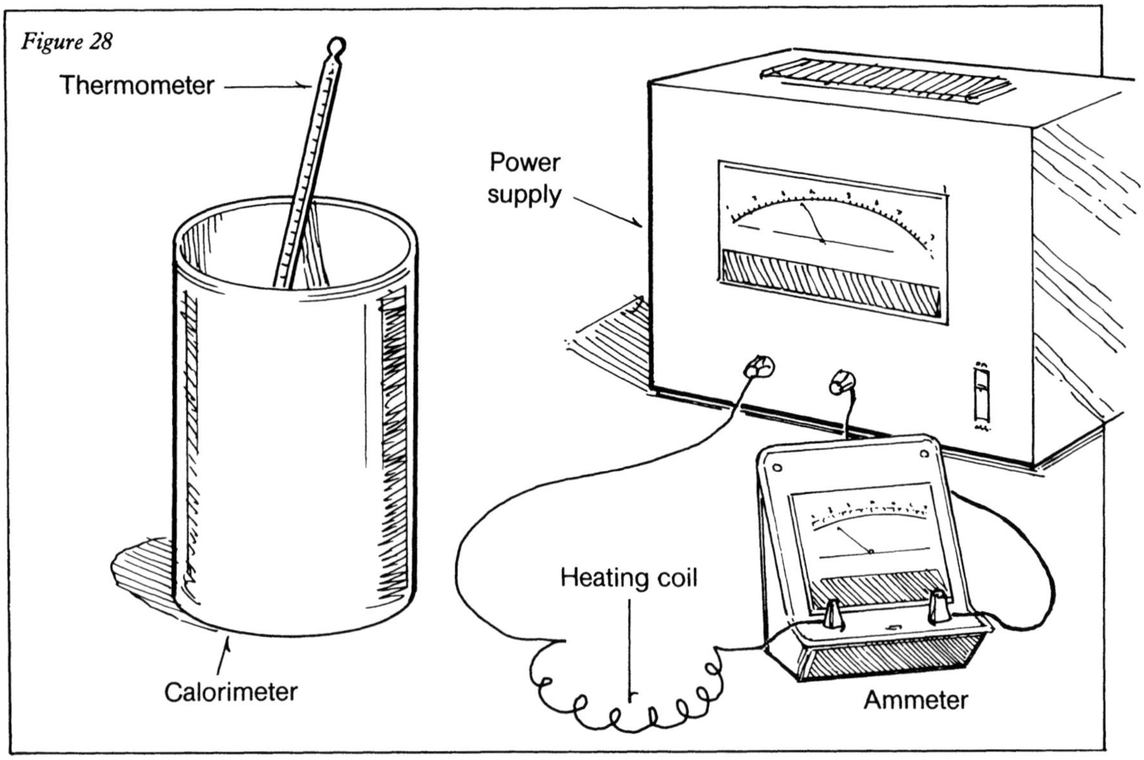

5. After you have waited a few seconds for the heat to leave the heating coil, stir the water and find the temperature of it using the thermometer as precisely as possible. Record this value.
6. Turn on the power for two minutes and record the voltage and current. Make sure the coil stays completely in the water.
7. After two minutes, turn off the power, wait a few seconds, stir the water, and then find the temperature again.
8. Repeat steps 6 and 7 twice more.
9. After you have gotten all the data you need, unplug all the wires.

(continued)

Lab #28

Name: ______________________

Electrical Equivalent of Heat *(continued)*

Date: ______________________

Cautions

1. Do not get water on the power supply, the ammeter, or the voltmeter. You can damage them or cause electrical shock.
2. Do not remove the coil from the water when current is going through it. It can easily cause burns. If it falls out accidently, turn off the power supply quickly.
3. Make sure that you stir the water and wait a minute before you measure the temperature.

Analysis

1. Find the mass of the water in the cup by subtracting the mass of the cup from the mass of the cup and water.
2. Using the specific heat of water (1 cal/gr-degree C) and the equation for heat gain, $Q = mc\Delta t$, find the heat gained by the water during each of your two-minute heating periods. This answer is in calories.
3. Find the amount of electrical energy given to the coil in two minutes using the equations: $P = VI$ and $W = Pt$
   which combine to give: $W = VIt$
   where W is the work in joules, V is the voltage in volts, I is the current in amps, and t is the time in seconds. This answer is in joules.
4. Convert the number of joules into calories using the conversion factor: 4.18 j/cal.

Conclusions

1. How does the amount of energy gained by the water compare to the energy given by the power supply? Explain.
2. Could any heat energy have been gained or lost from the system? Explain.
3. Are the errors which you find in your lab reasonable? Explain.
4. How could you improve this lab to reduce the error? Explain.

Quiz #28
Electrical Equivalent of Heat

Name: ______________________

Date: ______________________

1. In this lab electrical energy primarily is used to heat
   a. metal.
   b. glass.
   c. air.
   d. water.

2. What is the power output of a lamp which uses 3 amps of current at a voltage of 20 volts?
   a. 3 watts
   b. 7 watts
   c. 23 watts
   d. 60 watts

3. How many joules of energy does a 100-watt heating coil give off in two minutes?
   a. 50
   b. 200
   c. 2000
   d. 12,000

4. The biggest danger in this lab is
   a. losing your data by spilling your water.
   b. getting water on the power supply and getting a shock.
   c. not stirring the water enough to get the right temperature.
   d. hurting yourself with a voltmeter.

5. If 4.18 joules = 1 cal, how many calories are there if you have 41.8 joules?
   a. .1
   b. 1
   c. 10
   d. 100

Lab #28: Electrical Equivalent of Heat

Teacher Notes

Cautions

1. Do not get water on the power supply, the ammeter, or the voltmeter. You can damage them or cause electrical shock.
2. Do not remove the coil from the water when current is going through it. It can easily cause burns. If it falls out accidently, turn off the power supply quickly.
3. Make sure that you stir the water and wait a minute before you measure the temperature.

This lab is important because it shows the relationship of electrical energy to heat energy. Students should know this relationship for a complete understanding of conservation of energy. They should also know about this conversion to understand such things as electric bills. The lab reviews electrical as well as thermodynamics principles.

The combination of water and electricity in this lab is hazardous. If water gets into the power supply, students may get an electrical shock. Warn students about this and use electrical outlets with *ground fault interrupters* if at all possible.

The heating coil must stay in the water when the power supply is on. If it is out of the water, the heat from the coil will quickly make it hot to the touch and it can easily cause burns if left out of water for more than a few seconds.

The equipment can be purchased from physics supply houses. Science Kit has a suitable double-walled electric calorimeter for about $40. Less expensive ones are available for about $15. The least expensive ones can be made with Styrofoam® cups and resistors. A pair of Styrofoam® cups, one inside of the other, makes a good calorimeter. With this kind of calorimeter, the heat capacity can be ignored because it is so small. This write-up assumes a low-heat-capacity calorimeter so it ignores heat lost to it. This makes the lab easier to do and gives less error. The "heating coil" can be made using two or three 2-watt, 10-ohm resistors. The resistors are wound together in parallel to make a small resistance and attached to two 10-cm pieces of wire. The resistors will not last forever because the water will damage them. However, if you heat them briefly after use (just enough to drive the water out of them), they will last for a few years and are less expensive than other heating coils.

(continued)

Teacher Notes (continued)

You should have a 30–50-watt heating coil. This can be arranged by using the proper combination of voltage and resistance of the coil. The heating coil resistance generally is in the range of 3–5 ohms.

Possible Problems

1. Students may either not wait long enough or wait too long to take the temperature of the water. If so, the temperature may either not have risen to its highest value, or heat may have entered or left the system.
2. Students may spill the water and heating coil and lose their data.
3. Students may spill water on the lab benches and risk damaging the equipment or getting an electric shock. They must be careful when water and electricity are mixed.

Conclusions

1. Electrical energy should equal heat energy with less than 10% error.
2. Yes, if the water is colder than the room, heat can enter the water. If the water is warmer than the room, heat can leave. Also, evaporation is a problem if the temperature rises too much. Some energy must also go to heat up the calorimeter which does have some heat capacity.
3. Yes, some heat can be gained or lost. Heat is also needed to heat up the calorimeter.
4. Take into account the heat capacity of the calorimeter and make sure that the insulation allowed no heat to enter or leave the system.

LAB QUIZ ANSWERS

1. d
2. d
3. d
4. b
5. c

Lab #29
Specific Heat

Name: ______________________

Date: ______________________

In this lab you will investigate how heat is transferred from hot water to an object. You will add objects at room temperature to hot water and use the final temperature to find the specific heat of the objects. You will then add hot water to a coffee mug to find the specific heat of the mug. The equation for heat lost or gained is: $\Delta Q = mc\Delta t$ where ΔQ is the change in heat, m is the mass, c is the specific heat, and Δt is the change in temperature.

Materials

metal cup	thermometer
water	pieces of metal
hot plate	tongs
calorimeter	coffee mug
balance	

Procedure

Case 1:

1. Fill the metal cup half full of water and place it on the hot plate. Turn on the hot plate and heat the water to between 70 and 80 degrees Celsius (not boiling). Be careful not to burn yourself with the hot plate or spill water on it; this could cause electrical shock.
2. On a balance find the mass of your calorimeter and record it.
3. Fill a calorimeter one-quarter full with pieces of metal. Find the mass of the metal plus the calorimeter. Record this value.
4. The metal is initially at room temperature. Use the thermometer to find the temperature of the room as precisely as possible. Record this value.
5. When the water on the hot plate is between 70 and 80 degrees Celsius, use the tongs to fill the calorimeter half full. Be careful not to spill the hot water on you.
6. Put the thermometer into the calorimeter and read the temperature when the reading stays constant (see Figure 29). Record this value.
7. Find and record the mass of the cups, water, and metal.

(continued)

Lab #29
Specific Heat *(continued)*

Name: ______________________

Date: ______________________

Case 2:

8. Find and record the mass of the coffee mug.
9. Use the hot plate to heat enough water to fill the mug to a temperature of between 70 and 80 degrees Celsius. When the water on the hot plate is within that range, use the tongs to remove the container from the hot plate. Measure the temperature of the water precisely and record it. This is the initial water temperature.
10. Quickly pour the hot water into the mug, which is at room temperature. Use the thermometer to gently stir the water in the mug. Record the temperature when the reading stops dropping quickly. This is your final temperature. Record it.
11. Find the mass of the mug and water (make sure that you do not spill any water) and record this value.

> ***Caution***
>
> The hot plate and even the hot water can cause burns!

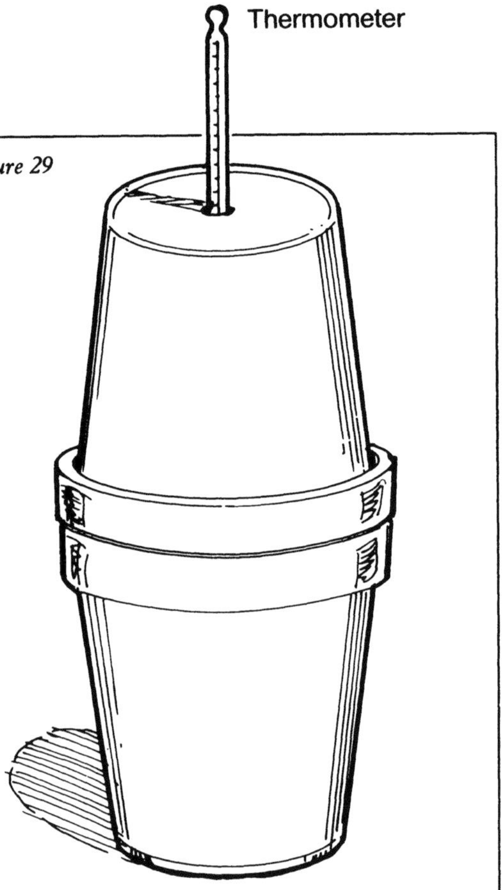

Figure 29

Styrofoam® cup calorimeter

Analysis

1. For Case 1, find the mass of the water and metal. Label these.
2. Use the final temperature to find the change in temperature of the water and of the metal.
3. The amount of heat gained by the metal equals the amount of heat lost by the water. Use the equation $\Delta Q = mc\Delta t$ to find the specific heat of the metal. The specific heat of water is 1 cal/gr-C.
4. Repeat steps 1–3 to find the specific heat of the coffee mug.

(continued)

Lab #29
Specific Heat *(continued)*

Name: ______________________

Date: ______________________

Conclusions

1. What is the value for the specific heat of the metal and what error do you think might exist in this value? Explain.
2. What is the value for the specific heat of the coffee mug and what error do you think might exist in this value? Explain.
3. What could add or take away heat from your cups or mug during this experiment which you have not taken into account? Are these significant sources of error?
4. How would it be possible to avoid having heat lost to the coffee mug when you add hot water? (This would keep the coffee hotter.)

Quiz #29
Specific Heat

Name: ______________________

Date: ______________________

1. Heat lost from one object is ______________ by another object.

   a. always lost
   b. always gained
   c. sometimes lost
   d. sometimes gained

2. If 20 calories of heat are added to 4 grams of water, what is the increase in the temperature of the water in degrees Celsius?

   a. 4
   b. 5
   c. 20
   d. 80

3. The mass of a cup is 4 gr. The mass of the cup with water is 208 gr. The mass of the water in grams is

   a. 204.
   b. 208.
   c. 212.
   d. 220.

4. Specific heat is defined as the number of calories required to

   a. heat a substance 1 degree Celsius.
   b. heat 1 gram of a substance.
   c. heat 1 gram of a substance 1 degree Celsius.
   d. cool a substance 1 degree Celsius.

5. In this lab you add water at a temperature of ______________ degrees Celsius to metal at room temperature.

   a. 0
   b. 30
   c. between 70 and 80
   d. 100

Lab #29: Specific Heat

Teacher Notes

Caution

The hot plate and even the hot water can cause burns!

In this lab students find the specific heat of a substance and then compare that value to the specific heat of a known object, a coffee mug. The lab is reasonably simple and does not require a lot of difficult measurements. The procedure has many steps, but they are all short.

The equipment for this lab includes a hot plate. Burners can also be used but the open flame makes them much more dangerous. The metal cup or can could be replaced with a pot although care must be taken to avoid hitting the handle and spilling the hot water. The heat capacity of the calorimeter can be ignored if it is made from a pair of Styrofoam® cups, one inside the other. A third cup can be made into a top by tearing off the top inch of the cup, punching a hole in the bottom for the thermometer, and sliding the cup upside-down into the other two.

The water temperature of between 70 and 80 degrees Celsius is chosen for two reasons. First, it will not burn as badly as boiling water and second, not as much evaporation and resulting heat loss will occur. Make sure that the students use water in this temperature range or lower. If their water boils, they can add cold water to lower the temperature.

Possible Problems

1. Students may discard the water before they have found its mass. This means that they have to start over.
2. Students may not be careful with the hot water and hot plate.
3. Students may be confused and have a hard time putting the proper numbers into the equations. Often, they will interchange numbers without knowing it.
4. Students may not measure the temperature precisely enough. They should be able to measure to tenths of degrees.

Conclusions

1. The specific heat for a few metals in cal/gr-C follows: aluminum, .22; brass, .092; copper, .092; lead, .031; tin, .054; zinc, .092. Student answers should be within about 10% of these values if they are careful with their measurements.

(continued)

Teacher Notes (continued)

2. The specific heat of ceramic materials is generally around .16 cal/gr-C although the student values may be high or low depending on the thickness of the mug and the amount of evaporation which takes place. If the mug is thick it will act as an insulator and not all of it will heat to the same temperature rapidly. This points out the difference between small pieces of metal which are good conductors and insulators like ceramics which are not.
3. Evaporation can take heat away.
4. Heat the mug by pouring hot water in before you add the hot coffee. Tea is sometimes prepared this way and it supposedly tastes better.

LAB QUIZ ANSWERS

1. b
2. b
3. a
4. c
5. c

Lab #30

Name: ______________________________

Heat of Fusion, or How Ice Cools a Drink

Date: ______________________________

In this lab you will investigate the heat of fusion and how it is used to efficiently cool drinks. You will add ice to warm water and use the temperature after the ice melts to calculate the heat of fusion. Because it takes energy to break the bonds when a substance changes from a solid to a liquid, when ice in a glass of water melts, the water around the ice loses heat. This energy is called the heat of fusion.

Materials	
calorimeter	ice
balance	thermometer
water	paper towels

Procedure

Case 1:

1. Find the mass of your calorimeter on a balance and record it.
2. Fill the calorimeter two-thirds full with warm water and then find the mass of the combination of water and calorimeter. Record this value.
3. Find the temperature of the water as precisely as possible and record it as the initial temperature (see Figure 30).
4. Take the amount of ice recommended by your teacher, use paper towels to dry it, and then add it to the water in the calorimeter.
5. Stir the ice and water gently with the thermometer until the ice melts. At this moment, find the temperature of the water. Record this temperature as final temperature 1.
6. Then place the calorimeter back on the balance to find the total mass. Record this value.

Case 2:

7. Now add the amount of ice suggested by your teacher to the water. The purpose is to cool the water down to zero degrees in much the same way soda is cooled. Measure the temperature of the water with the thermometer and *gently* stir the water and ice until the temperature stops falling. Then remove any ice which has not melted. Record the temperature as the final temperature.

(continued)

Name: ______________________

Date: ______________________

8. Find and record the mass of the calorimeter with the water and melted ice.

Figure 30

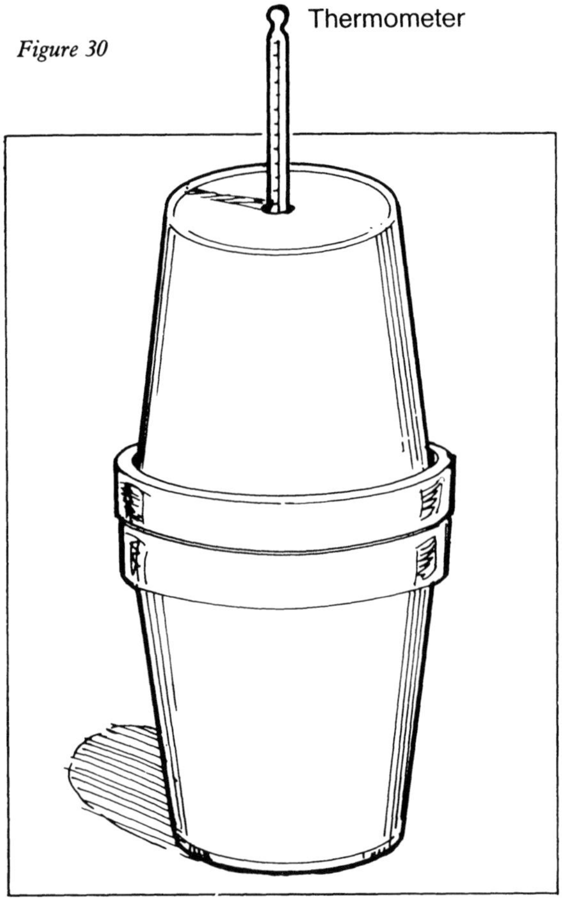

Styrofoam® cup calorimeter

Cautions

1. Make sure that you find the mass or temperature which is required before going on to the next step. For instance, if you add ice before you have measured the temperature, you will have to start over.
2. Measure the temperature as precisely as possible. Ideally you should measure to tenths of a degree.

Analysis

1. In Case 1, find the mass of the water in the calorimeter and the mass of the ice added by using the masses which you measured.
2. Using the final temperature, find the change in temperature of the water and the water that came from melted ice. The melted ice starts at zero degrees.
3. Use the equation for heat lost and heat gained to find the heat of fusion. You have all the variables except for the heat of fusion.
4. In Case 2, find the total mass of ice which melted to bring the water to near zero degrees. (This includes the ice in Case 1.)
5. Use the equation for heat lost and heat gained to find the heat of fusion.

(continued)

Lab #30

Name: ______________________

Heat of Fusion, or How Ice Cools a Drink *(continued)*

Date: ______________________

Conclusions

1. The value for the heat of fusion of ice is actually 80 cal/gr. How do your values compare to this? Which one is closer? Why might that be?
2. Do your measurement errors explain the difference between your value and the accepted value? Explain.
3. What sources of error other than measurement error may affect the outcome of the experiment? How large are these sources?
4. If ice had been added to a soda, what effect on the flavor of the soda would the melted ice water have had? What would have happened if you had added ice to soda at a much higher temperature, say one left in the sun? What would happen if you held onto the drink for a long time and more ice melted?

Quiz #30

Heat of Fusion, or How Ice Cools a Drink

Name: ______________________________

Date: ______________________________

1. If a small amount of ice at 0 degrees Celsius is added to a large amount of water, heat is given to the ice as it
   a. melts.
   b. warms.
   c. melts and then warms.
   d. warms and then melts.

2. If 40 calories of heat are added to 5 grams of water, what is the increase in the temperature of the water in degrees Celsius?
   a. 5
   b. 8
   c. 45
   d. 200

3. The mass of ice added to the water in the calorimeter is found by subtracting the mass of the __________________ from the total mass.
   a. water
   b. water and calorimeter
   c. calorimeter
   d. water, calorimeter, and ice

4. Heat of fusion is defined as the number of calories required to
   a. heat a substance 1 degree Celsius.
   b. heat 1 gram of a substance 1 degree Celsius.
   c. cool 1 gram of a substance 1 degree Celsius.
   d. melt 1 gram of a substance.

5. When a substance melts,
   a. energy is used to break bonds.
   b. energy is used to raise the temperature.
   c. energy is lost as the temperature drops.
   d. no energy is exchanged.

Lab #30: Heat of Fusion, or How Ice Cools a Drink

Teacher Notes

Cautions

1. Make sure that you find the mass or temperature which is required before going on to the next step. For instance, if you add ice before you have measured the temperature, you will have to start over.
2. Measure the temperature as precisely as possible. Ideally you should measure to tenths of a degree.

In this lab students experiment with the heat of fusion and verify the value for ice. Two methods are used to give the students more practice with the calculations and also to try to make sure that at least one of the results will be precise. Experimental technique may introduce errors in one of the methods and not the other. Students may also learn the effect of ice on the flavor of a drink. Generally, the results are good if students avoid error by being careful with their measurements. Calculation errors also seem to creep in because students do not subtract the mass of the cup from the water, etc.

Any calorimeter can be used, but Styrofoam® cups work well, particularly because they have such a low heat capacity. The lab must be slightly modified to include the heat gained or lost by the calorimeter cup if other calorimeters are used. With standard metal-cup calorimeters, the heat capacity of the cup needs to be accounted for in the equation for heat lost and heat gained. (This heat gained or lost by the cup can be ignored although the accuracy of the results may suffer.) The Styrofoam®-cup calorimeter is constructed using two cups, one inside the other. A third cup can have the top inch torn off and a hole punched at the center of the bottom. When inverted, it makes a good top for the calorimeter and the hole can be used to insert the thermometer. This inexpensive calorimeter produces good results and actually makes the calculations easier.

In Case 1, students should add the amount of ice which will lower their water temperature to about the same amount below room temperature as the warm water was above room temperature. Because the volume of calorimeters varies, you should experiment to find the number of ice cubes required. The ice should also be out of the freezer for about 10 minutes so that it is at zero degrees Celsius. The warm water should not be warmer than about 30 degrees Celsius or it will overflow in Case 2. Also, when students add ice in Case 2, make sure that water does not overflow when it is added or removed. The correct amount of ice can be found by experimentation.

(continued)

Teacher Notes (continued)

Possible Problems

1. Students may neglect to make a measurement before moving on to the next step. They must start over.
2. Students may get lost in the calculation and put the wrong numbers into the equation.

Conclusions

1. Generally, the results should be within about 10% of the given value. Both values should be about the same. Differences may exist depending on the lab technique.
2. The measurement errors should explain the differences.
3. Water lost from or added to the cup will cause an error in the mass. The size of this error depends on how carefully the experiment is run. Errors from the thermometers should be very small. Heat gained or lost from the system should be only a small problem.
4. The soda will be diluted and not as strongly flavored. If you start with soda at a higher temperature, it would be more dilute. If you hold onto the drink for a long time, it will become still more dilute. This is why the last swallow of a soda may not taste like the soda at all.

LAB QUIZ ANSWERS

1. c
2. b
3. b
4. d
5. a

Lab #31
Solar Energy

Name: ______________________

Date: ______________________

In this lab you will investigate the rate at which the sun gives heat to containers of water left in sunlight. The energy which the water absorbs is used to find the solar constant. The solar constant is the amount of heat energy which the sun gives to each square centimeter of the earth each minute. The energy reaching the upper atmosphere is about 2 $\frac{\text{Cal}}{\text{cm}^2 \cdot \text{min}}$.

Materials

- balance
- flat-bottomed containers (one black on the bottom and the other white)
- water
- thermometer
- aluminum foil
- cardboard
- meter stick
- timer
- protractor

Procedure

1. With a balance, find the mass of each container. Record this value along with the container number which will allow you to identify the container later on.
2. Fill each container to within 2 centimeters of the top. Then find the mass of the water plus the container. Record this value along with the container number.
3. With a thermometer, measure the temperature of the water as precisely as possible and record it to the nearest tenth of a degree if possible.
4. Bend a piece of aluminum foil so that it covers the outside of one side of the container. For the black-bottomed container, have the foil on the side opposite the black side. You can wrap the foil slightly over the top of the container to hold the foil in place. Then place a piece of cardboard over each container so that you can prevent sunlight from entering the container until you start the experiment.
5. Carefully take the containers to a sunny spot. Make sure the side with the aluminum foil is the side toward the sun (see Figure 31).
6. Remove the cardboard and time for a five-minute interval. After five minutes, measure the temperature of the water in each container as precisely as possible and record it. Repeat the heating for another five minutes and measure the temperature again.

(continued)

Name: ______________________________

Date: ______________________________

7. If at any time the sun goes behind a cloud, record the time that has passed since you removed the cardboard. Also record the temperature of the water.
8. Measure the length and width of the container's top. Record these values.

Figure 31

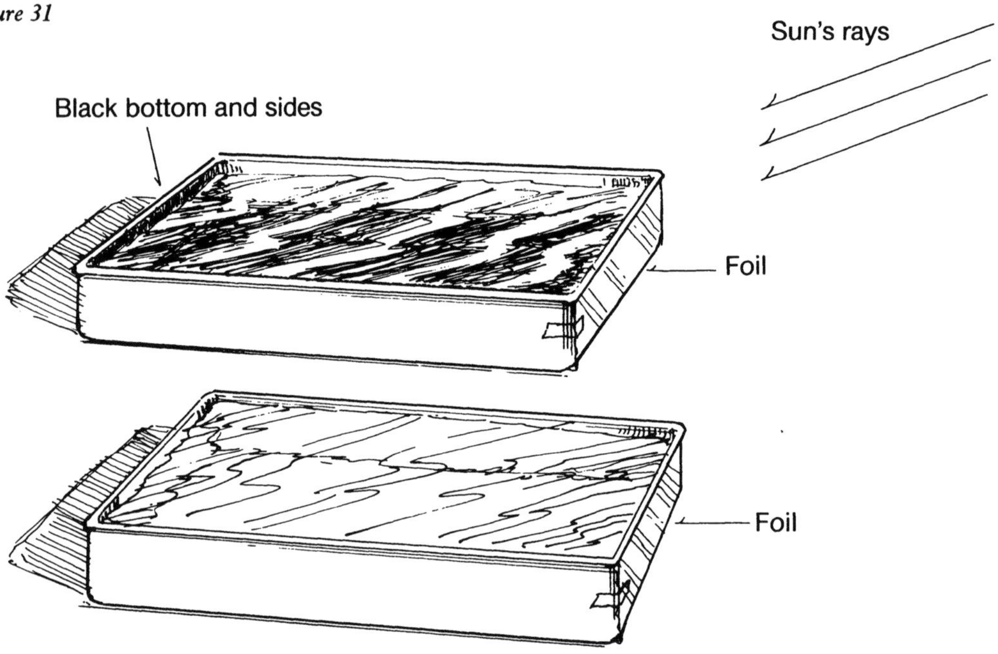

9. Use the meter stick to find the angle at which the sun strikes the ground. Move the meter stick so it is parallel to the sun's rays and causes the smallest shadow. Use a protractor to measure this angle.

Analysis

1. Find the amount of heat gained by the water in the first and second five-minute periods for each container using the equation: $\Delta Q = mc\Delta T$ where ΔQ is the heat gained, m is the mass of the water, c is the specific heat of water, and ΔT is the change in temperature of the water.
2. Find the surface area of the top of the containers.
3. Find the number of calories per minute per square centimeter the sun gave to the water for each period for each container.

(continued)

Name: ______________________________

Date: ______________________________

4. The amount of solar energy striking the top of the atmosphere is 2.0 cal per square centimeter per minute, the solar constant. This occurs if the sun's rays are perpendicular to the surface collecting the energy. At other angles, the solar constant must be multiplied by the sine of the angle at which the rays strike. Multiply the solar constant by the sine of the angle at which the sun's rays strike your container.

Conclusions

1. How does the energy collected per minute in your black-bottomed collector compare to the value you calculated in step 4?
2. How does the energy collected per minute in your light-colored collector compare to the value you calculated in step 4? Why is there a difference as compared to the black-bottomed container?
3. What factors could cause the difference between your values and the energy reaching the top of the atmosphere?
4. Where can energy be lost in your experiment? How large do you think the losses are? Explain.
5. What purpose does the aluminum foil serve in this experiment?

Quiz #31
Solar Energy

Name: ______________________________

Date: ______________________________

1. In this lab, the sun will heat __________________ and the change in temperature will be noted.
   a. water
   b. rocks
   c. dirt
   d. air
2. The solar constant is equal to the amount of energy
   a. each square centimeter of the earth gets.
   b. each square centimeter of the earth gets per minute.
   c. the earth gets over the period of the year.
   d. absorbed by an object each second.
3. In this lab the sun heats water in two containers; one
   a. is metal and the other is plastic.
   b. is big and the other is small.
   c. has a mote in it and the other doesn't.
   d. has a black bottom and the other has a light bottom.
4. You must try to measure the temperature of the water to the nearest
   a. degree.
   b. tenth of a degree.
   c. 10 degrees.
   d. It does not matter how close it is measured.
5. Can you continue your experiment if the sun goes behind a cloud and get good results?
   a. yes
   b. no
   c. sometimes
   d. maybe

Lab #31: Solar Energy

Teacher Notes

In this lab students look at solar heating and the solar constant. The importance of this constant for the world is great and if students have run a simple experiment to find its value, then they may have a greater appreciation for solar energy. The lab cannot be done on a cloudy day. However, because it is an easy and quick lab, it can be done on the spur of the moment when the sun is out. You will also get better results if you do not do the lab in the winter when the sun's rays are less direct. It should also be done around noon if possible.

For this experiment you may use food-storage containers such as sandwich containers. The sides of the containers should be two inches high or less. Otherwise, the large amount of water causes only a small change in temperature. The inside of the bottom and one side of the container should be painted black. (Pieces of black enameled metal can also be cut to fit in the bottom and up the side if they are available.) The containers should be numbered so that when they are left outdoors students can identify them.

The lab has sources of error, such as heat lost through the container's walls and evaporation, clouds moving in front of the sun, energy reflected from the water's surface, and energy absorbed by the atmosphere.

You can also investigate a number of different aspects of solar heating by having different lab groups use different combinations of heating systems. Some variations include plastic wrap over the top, glass resting on the top, two layers of glass as in a double-glazed window, and glass at different angles to the sun placed over the container. The different angles will cause differing amounts of heat energy to enter the container because different amounts of heat are reflected from the surface.

Possible Problems

1. When the sun goes behind a cloud, make sure that students mark the time, measure the temperature, and place the cardboard over the container until the sun comes out again. They will then have to calculate the solar constant using a different amount of time.
2. If you cannot go outside to do the measurements, glass in windows and screens on windows will affect your results. This will reduce the heat gained by the water but can be instructive as to how much heat energy enters a house through a window.

(continued)

Teacher Notes (continued)

Conclusions

1. The heat gained by the container is less than the solar constant by about 10% with good conditions and no evaporation. The atmosphere absorbs about 5% of the heat. The rest is related to collector loss.
2. The black absorbs more heat because it absorbs rather than reflects most of the heat energy.
3. Energy absorbed by the atmosphere or reflected by high clouds.
4. Heat is lost through the walls of the container, reflected from the water surface, and absorbed by the container itself.
5. The aluminum foil prevents heat from entering the side of the container.

LAB QUIZ ANSWERS

1. a
2. b
3. d
4. b
5. b

Lab #32
Heat Loss and Insulation

Name: ______________________

Date: ______________________

In this lab you will investigate how heat is lost through different containers. You will investigate the insulating properties of a number of different containers and then see if the rates of heat loss agree with the equation for heat loss.

The equation for heat loss per unit time is

$$\Delta Q/\Delta t = kS\Delta T/d$$

where $\Delta Q/\Delta t$ is the heat lost per unit time, k is a constant related to the material (k is small for insulators and large for conductors), S is the surface area, ΔT is the change in temperature, and d is the thickness of the insulating layer.

Materials

metal conatiners	paper
glass beaker	2 thermometers
graduated cylinder	timer
Styrofoam® cups	water
marker	

Procedure

1. Set up a series of six container systems in the following ways:
   a. one metal container
   b. one glass beaker
   c. one graduated cylinder
   d. a single Styrofoam® cup open at the top
   e. a single Styrofoam® cup, with a top made out of a Styrofoam® cup.

      (The top is made by cutting off about 2 cm of the top edge of a cup and poking a hole in the bottom for a thermometer to go through. This cup is then turned upside down and slid into the top of the other cup.)

   f. two Styrofoam® cups, one inside the other and with a top similar to the one described in *e*.

2. Fill one of the containers about three-quarters full. Mark the water level on the container, and then pour the water out of the first container into one of the other containers and mark the water level on the inside of the container.

(continued)

Name: ______________________

Date: ______________________

Repeat the process until all of the containers have a mark on them. This mark will be used to make sure that each container has the same amount of water added to it. Empty all containers.

3. Now place a piece of paper under each container and label each piece with the name of the container which it is under. These sheets will serve as your data tables for recording time and temperature (see Figure 32). Each lab partner will copy the data from these sheets into his or her own data table at the end of the lab.

Figure 32

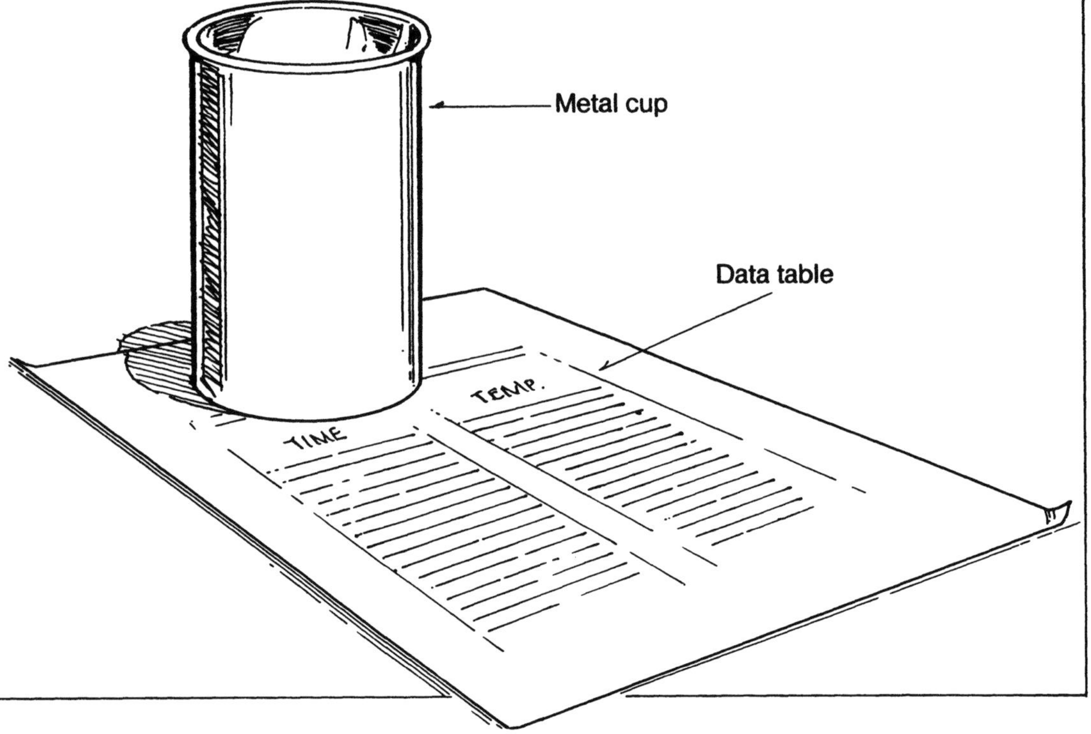

4. Now fill each container with hot water to the mark which you drew in step 2. Record the time and then place a thermometer into the hot water which is coming from the faucet to get the temperature of the water. Make sure you leave the thermometer in the water until the reading stops changing. Write this temperature on each sheet for time zero minutes.
5. Place a thermometer in the metal container and one in the glass beaker. Leave the thermometers in the containers for one minute or until the reading is not changing. Then read the temperatures, record the time and temperature for each

(continued)

Name: ____________________

Date: ____________________

reading on the sheet under the container, and put the thermometer into another container or beaker. The idea is to get the temperature at various times for all six containers. If one temperature does not seem to be changing much, you do not have to measure it as often as the more quickly changing ones.

6. Make sure that each lab partner has a copy of the data before leaving the lab.

Cautions

1. Be careful not to allow the thermometer to fall to the floor.
2. Make sure that you do not tip over the containers as you measure the temperature. The thermometer may make the containers tip easily if you do not hold them.
3. Make sure that the reading on the thermometer is not changing when you read it.

Analysis

1. Draw a set of axes with temperature on the vertical axis and time on the horizontal. Make sure that the temperature axis has enough room to go from the highest temperature you measured to the lowest. The time axis should allow you to plot points for the entire duration of the experiment.
2. Using different symbols for each container (squares, circles, triangles, etc.), draw a graph of temperature vs. time for each container. Draw a smooth curve through each of these sets of points. You will have six different lines on the same graph. You may find it helpful to use different colors to show the different lines.

Conclusions

1. Which of the graphs shows the temperature dropping the fastest? Does the equation predict this? Explain.
2. Which of the graphs shows the temperature dropping the slowest? Does the equation predict this? Explain.
3. What difference was there, if any, when a top was used on the containers? Why might this have an effect on the cooling rate?

(continued)

Lab #32
Heat Loss and Insulation *(continued)*

Name: ______________________________

Date: ______________________________

4. How did a different thickness of insulation affect the cooling? Explain.
5. Looking at the cooling curve for the metal container, how does the *rate of change* of temperature change with time? Does the equation predict this? Explain.
6. The graduated cylinder had a different surface area. Did the temperature change occur as you would expect from the equation?
7. How does the equation relate to these curves? Which of the variables were changed in the lab and what effect did they have on the cooling rates? Explain.
8. What do you think the temperature of all of the containers would be after a few hours time? Why do you choose this temperature?

Quiz #32
Heat Loss and Insulation

Name: ____________________

Date: ____________________

1. In this lab you will measure the temperature
   a. of the room at various times.
   b. of each container of water once.
   c. of each container of water a number of times.
   d. of only one container of water at various times.

2. The equation for heat loss shows that if the temperature difference between two objects is great, the heat loss from one to the other is
   a. great.
   b. sometimes great and sometimes small.
   c. small.
   d. It can't always be determined.

3. The equation for heat loss shows that if the thickness of the insulation is large, the heat loss from one to the other is
   a. great.
   b. sometimes great and sometimes small.
   c. small.
   d. It can't always be determined.

4. You must be careful of the thermometers during the lab because
   a. they get very hot and can burn you.
   b. they get very cold and can freeze you.
   c. they are very sticky.
   d. if they drop to the floor they may break.

5. The paper placed under each of the containers is used for
   a. mopping up water.
   b. holding the containers steady.
   c. recording data.
   d. insulation.

Lab #32: Heat Loss and Insulation

Teacher Notes

Cautions

1. Be careful not to allow the thermometer to fall to the floor.
2. Make sure that you do not tip over the containers as you measure the temperature. The thermometer may make the containers tip easily if you do not hold them.
3. Make sure that the reading on the thermometer is not changing when you read it.

In this lab students gain a better understanding of the heat-loss equation. They also learn a bit more about how things cool and about insulation. The lab gets a bit hectic because students are constantly taking temperatures, but the data generally turn out well.

The apparatus is simple. The metal container may be any can but the glass beaker should be Pyrex®. The container and beaker should be about the same size as the Styrofoam® cups so that they hold about the same amount of water. The graduated cylinder must hold the same amount of water as the cups. The water should be as hot as the hot-water tap can produce. Let it run for a while so that it is at constant temperature.

Possible Problems

1. Students may not record data carefully and may get the data from one container mixed up with another.
2. Thermometers may fall to the floor. Warn students about this.

Conclusions

1. The metal container should lose heat fastest because it has a high value of *k* and a small thickness.
2. The slowest cooling should be the double container with the top. It has a small *k*, large *d*, and has no evaporation from the top.
3. The lid should prevent evaporation and insulate the top to prevent convection at the water's surface.

(continued)

Teacher Notes (continued)

4. The thicker insulation should reduce the cooling rates. The difference may not actually be a factor of two.
5. The rate of change of temperature gets smaller and smaller because the temperature difference becomes smaller and smaller.
6. The larger surface area should cause faster cooling unless the graduated cylinder is very thick and then the size of the area exposed to evaporation may mean that it cools more slowly.
7. The equation predicts the rates of cooling shown here with tests of type of insulating material (k), surface area (S), temperature difference (T), and thickness of insulation (d).
8. The temperature will approach room temperature. When the temperature difference is zero, the heat loss is zero. In theory you never reach a point where the containers reach room temperature but the difference becomes hard to measure or the room's temperature fluctuates.

LAB QUIZ ANSWERS

1. c
2. a
3. c
4. d
5. c

Lab #33
Expansion of Water

Name: ______________________

Date: ______________________

In this lab you will investigate the expansion of water by observing its change in volume with changes in temperature. You will attach a container of water to a thin tube. As the water expands in the container, you will see the small change in volume in the thin tube. You will observe changes in volume of water as the temperature changes from 0 degrees Celsius to near 30 degrees Celsius.

Materials

thermometer	container for water
two-holed stopper	large container
glass tube	ice
glycerine	water
ruler	

Procedure

1. Insert the thermometer into one of the holes of the two-holed stopper and then insert the glass tube in the other. Glycerine will help the thermometer and tube to slide into the holes easily.
2. Fill the container to the top with water.
3. Holding the container upright, insert the stopper into the mouth of the container. Make sure that no air is in the container. Water should fill the tube. Attach a small ruler to the thermometer and tube so you can measure the position of the water in the tube. See Figure 33.
4. Place the container, thermometer, and glass tube (which will be called the "apparatus") in a container of ice water. When the temperature on the thermometer reaches zero degrees, remove the apparatus. Move the stopper out so that only about 1 cm of water shows beyond the stopper. Return the apparatus to the ice water.
5. Measure precisely (to the nearest .001 m) the distance the water in the tube is from the stopper. Record this value.
6. Remove the apparatus from the ice and allow the water to warm up to 1 degree Celsius. Hold the container in your hand to add heat, and gently shake the apparatus as you warm it to insure uniform heating. When the temperature has reached 1 degree, measure the distance which the liquid is from the stopper.

(continued)

Name: ______________________

Date: ______________________

7. Increase the temperature slowly, in 1-degree amounts, and record the temperature and distance the water is from the stopper. Do this until you have raised the temperature as high as you can by holding the container in your hand. This should be about 30 degrees Celsius. Stop heating the water if it expands out of the glass tube.

Figure 33

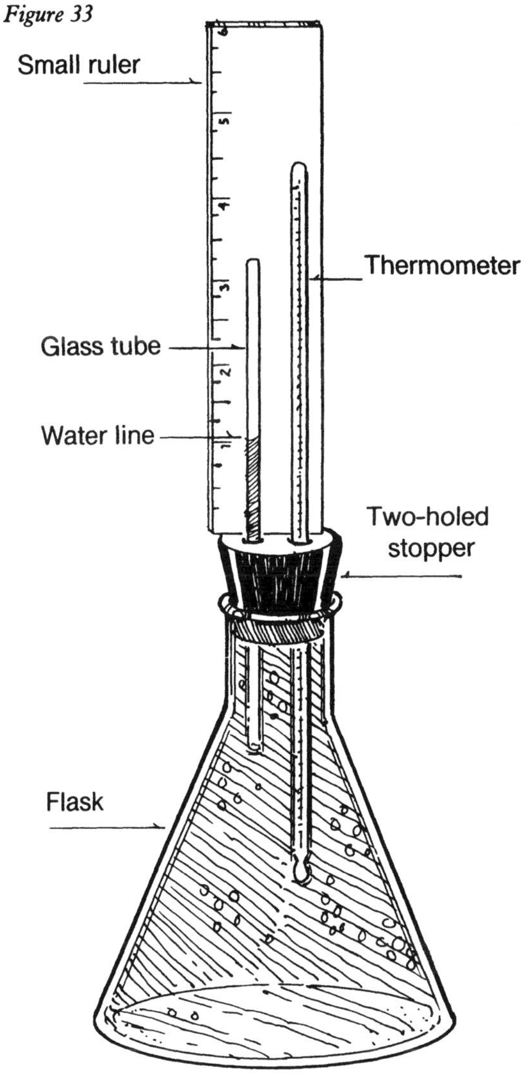

Cautions

1. The thermometer and glass tube can break when you insert them into the stopper. If they do, the glass can cause serious harm. Handle the apparatus carefully.
2. Make sure that you do not drop the apparatus for it may break and scatter small pieces of glass.

Analysis

Draw a graph of temperature vs. position of the water in the tube. Draw a smooth curve through these points.

Conclusions

1. When the water expands, it pushes water up the tube. When the water contracts, it pulls water down the tube. At what point has the water contracted to its smallest volume? Does this make sense according to what you know about water?

(continued)

Lab #33
Expansion of Water *(continued)*

Name: ______________________

Date: ______________________

2. Does the motion of the water up and down the tube surprise you? Where else have you seen liquids change in volume as a response to temperature? Explain.
3. At what point would the density of the water in the container be the largest?

Quiz #33
Expansion of Water

Name: ______________________________

Date: ______________________________

1. The thin tube attached to the larger container of water allows
   a. the water to expand more easily.
   b. the expansion of water to be seen more easily.
   c. the motion of water in the container.
   c. water to flow out of the container.

2. The temperatures which you will be working with in this lab will be ____________________ degrees Celsius.
   a. between –30 and 0
   b. between 0 and 30
   c. between 30 and 60
   d. above 60

3. Heat is added to the water by
   a. a burner flame.
   b. a special heater.
   c. a hot plate.
   d. your hands.

4. When most liquids are heated they
   a. expand.
   b. expand and then contract.
   c. contract.
   d. contract and than expand.

5. You will measure the temperature of the water in the container and the
   a. size of the oscillation of the water in the container.
   b. time to change the temperature of the water one degree.
   c. mass of water in the tube.
   d. displacement of water in the tube.

Lab #33: Expansion of Water

Teacher Notes

Cautions

1. The thermometer and glass tube can break when you insert them into the stopper. If they do, the glass can cause serious harm. Handle the apparatus carefully.
2. Make sure that you do not drop the apparatus for it may break and scatter small pieces of glass.

In this lab students observe how water expands. The expansion of water around 4 degrees Celsius is particularly important because at that temperature and below it, water is abnormal. This abnormality is important for students to investigate.

The apparatus that is used must show a motion of water up and down the tube large enough to measure. Following are guidelines for the selection of sizes. The container should have a mouth large enough to accept a two-holed stopper and should be about 250 ml in volume. The tubing should be standard glass tubing about 30 cm long. You should make sure that you experiment with your apparatus before you present it to the students.

It may help if you construct the apparatus before the lab. Putting thermometers and glass tubing through the stopper holes is not difficult, but use of a lubricant will probably prevent the breaking of a thermometer or glass tube. It is important to use a lubricant such as glycerine which doesn't dry out so that the thermometers and tubing can be removed easily. Unless you want to keep the apparatus as permanent equipment, you should take it apart after the lab is done. It will prevent sticking and make the process easier.

Cooling the containers to zero degrees may be time-consuming unless you use one of the following methods: 1) If you make the apparatus beforehand, you can place the containers in an ice bath for an hour or more before the lab. The temperature will approach zero but probably will not reach it. 2) You can use salt to lower the temperature of the ice-water mixture. This will cool the containers to zero. If ice starts forming in the container, expansion will occur and confuse the students. The change in volume as ice freezes or melts makes adding ice directly to the containers a poor method for cooling the water. Unless all the ice melts before the readings are taken, melting ice will affect the data.

This lab can be made more numeric if you choose to find the coefficient of expansion (which is 0 at 4 degrees and 2.1×10^{-4} at 20 degrees). In order to do this,

(continued)

you must find the volume of the container by using the mass of water enclosed and the cross sectional area of the tube. This can be done with a micrometer. One thing which has not been taken into account in this lab is the fact that the glass expands as well as the water. This effect is smaller by a factor of 100 than the expansion of the water so you still can get some reasonable data. However, if you are going to try to calculate the coefficient of expansion, you might want to take this other expansion into account.

Possible Problems

1. Students may allow air into the container or they may not bring the water in the tube to 1 cm above the stopper.
2. Students may not measure to the precision which is required to find the small changes which occur around 4 degrees Celsius.

Conclusions

1. The smallest volume should be at 4 degrees but the volume doesn't change much around that temperature.
2. This motion should surprise most students because it is much larger than they realize.
3. The density is largest at 4 degrees when the volume is smallest.

LAB QUIZ ANSWERS

1. b
2. b
3. d
4. a
5. d

Lab #34
Boyle's Law and Compressed Air

Name: ______________________

Date: ______________________

In this lab you will investigate the compressibility of air under pressure. Air in a tube is compressed as weights are attached to a piston. As the pressure increases, the volume should decrease according to the equation PV = constant. This relationship between pressure and volume is called Boyle's Law because it was first investigated by Robert Boyle in the late seventeenth century.

Materials

air cylinder and piston	weights
support stand	meter stick
clamp	

Procedure

1. Attach the air cylinder and piston to the support with the clamp. Make sure that it is securely attached but that the clamp does not squeeze the cylinder too much.
2. Measure the distance between the lower end of the piston and the mark which shows the largest volume of the cylinder. Record this value.
3. Record the largest volume for the cylinder in milliliters or cubic centimeters.
4. Open the valve at the bottom of the cylinder to allow air to go in and out as the piston moves. Insert the piston so the bottom of it is at the mark on the scale which has the largest volume.
5. Close the valve at the bottom of the piston and then make sure that no air moves in or out when you exert force on the piston.
6. Now hang a mass of the size specified by your teacher on the hanger (see Figure 34). Record the mass and the position of the piston in the cylinder. Measure the position to the nearest tenth of a division.
7. Repeat the process adding mass at the rate suggested by your teacher, each time recording the mass and the position of the cylinder. Keep adding mass until you have reached the maximum amount given by your teacher.

(continued)

Lab #34
Boyle's Law and Compressed Air *(continued)*

Name: ____________________

Date: ____________________

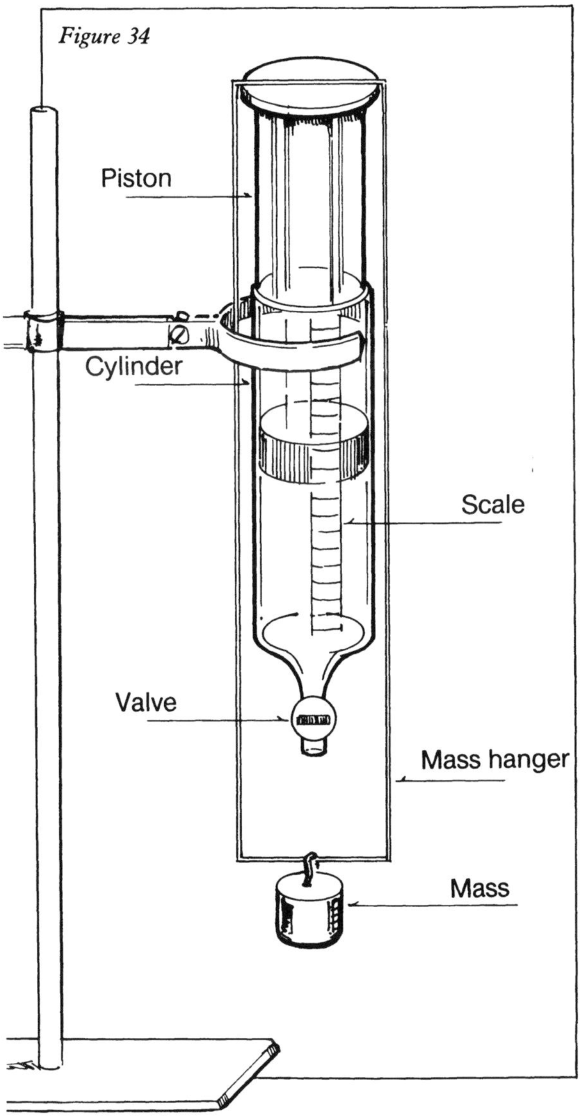

Cautions

1. Make sure that you do not exceed the amount of mass which your teacher suggests. This could break the apparatus and if the masses fall, they may injure you.
2. Make sure that the apparatus is mounted vertically or the large forces may cause it to tip or bend.

Analysis

1. The volume of the cylinder is related to the area of one end of the cylinder times the length. Find the area of the end of the cylinder by dividing the volume in cubic centimeters by the length in centimeters. Your answer will be in square centimeters. (1 milliliter equals 1 cubic centimeter.) The piston is one end of the cylinder so its size equals the area of the end of the cylinder.
2. Pressure is exerted on the cylinder when a force is exerted on the piston. Pressure is force per unit area so to find the pressure in the cylinder, the gravitational force exerted by the mass is divided by the area. Find the gravitational force exerted by each of the masses which you used and then find the pressure by dividing this force by the area of the cylinder.
3. Plot a graph of the pressure vs. volume with volume on the horizontal axis.
4. Find the values for 1/volume for each of the volumes which you used and then plot a graph of pressure vs. 1/volume with 1/volume on the horizontal axis.

(continued)

Lab #34

Boyle's Law and Compressed Air ***(continued)***

Name: ______________________

Date: ______________________

Conclusions

1. Describe the graph of pressure vs. volume. What does it look like? Should it look like this?
2. Describe the graph of pressure vs. 1/volume. What does it look like? Why should it look like this?
3. Atmospheric pressure at the surface of the earth is about 10 N/cm^2. The forces exerted on the piston by the masses add to this pressure. When the pressure from the masses equals 10 N/cm^2, the sum of the pressures is 20 N/cm^2 ($10 + 10 = 20$) or twice the original pressure. Using your graph, what is the volume when the pressure is twice as great? Does this agree with what the Boyle's Law equation relating pressure and volume says?
4. What would the pressure be to make the volume one third as big according to the equation? Does your data support this?
5. At what point would the volume be reduced to zero? Explain. Does your data support this?
6. This lab has not taken into account the change in temperature of the gas when it is compressed. Given the way each mass was added, do you think that temperature changes are a big problem? What effect would temperature have?

Quiz #34
Boyle's Law and Compressed Air

Name: ______________________________

Date: ______________________________

1. Boyle's Law gives the relationship of
   a. temperature and pressure.
   b. temperature and volume.
   c. pressure and volume.
   d. pressure, volume, and temperature.

2. As the pressure of a gas increases, the volume of the gas normally
   a. increases.
   b. stays the same.
   c. decreases.
   d. increases and then decreases.

3. In this lab pressure is increased in the cylinder by
   a. pushing with your hand.
   b. using a pump.
   c. using a simple lever.
   d. adding mass to a piston.

4. A gas has a volume of 4 liters. If the pressure is doubled, what is the new volume in liters?
   a. 1
   b. 2
   c. 4
   d. 8

5. If the apparatus is not mounted vertically,
   a. the large forces can break or bend the piston.
   b. the air will not be compressed.
   c. Boyle's Law cannot be proven with any accuracy.
   d. mass cannot be placed on the piston.

Lab #34: Boyle's Law and Compressed Air

Teacher Notes

Cautions

1. Make sure that you do not exceed the amount of mass which your teacher suggests. This could break the apparatus and if the masses fall, they may injure you.
2. Make sure that the apparatus is mounted vertically or the large forces may cause it to tip or bend.

In this lab students investigate Boyle's Law ($PV = k$). Students see how the change in pressure causes a change in volume and then relate this to the equation. Students draw two graphs, one being P vs. V and the other being P vs. $1/V$. They investigate the shape of each graph and explain why they are shaped as they are. Students discuss temperature effects although they do not specifically investigate them.

The equipment used for the lab can be purchased from Central Scientific Company for about $15 and is called the Boyle's Law Apparatus. It is particularly handy because of the valve at the bottom of the cylinder and also because mass is hung on the piston rather than balanced on the top. A regular piston (syringe) without the valve can be used but the scale on the cylinder may have to be extended and the balancing of masses on top may cause problems in the lab.

Before the lab, determine the maximum amount of mass that can be hung on the piston. Because the amount of mass used is quite large, you may have to resort to masses such as bricks rather than the standard lab masses. If you have to use nonstandard masses, you will have to find ways to attach them to the apparatus.

Possible Problems

1. If the cylinder is not mounted vertically, the large forces may cause the apparatus to tip or bend.
2. Some students are not familiar with the shape of a hyperbola so they are not able to discuss the shape of the P vs. V graph. They may also not understand the significance of plotting $1/V$ which should give a straight line.
3. With time, the pistons may start to stick. If this happens, the students should be directed to very gently bounce the mass on the piston so that they can find the equilibrium value. A silicon spray might also help but make sure that the spray does not affect the plastic cylinder or rubber on the piston.

(continued)

Teacher Notes (continued)

Conclusions

1. The graph should be a hyperbola and agree with the shape predicted by the equation.
2. The graph should be a straight line and agree with what the plot of P vs. $1/V$ should be.
3. With an extra atmosphere of pressure, the volume should be half.
4. With two extra atmospheres of pressure, for a total of three, the volume should be one third.
5. The volume will only reach zero with infinite pressure which is of course not possible. The data should support this because the graph is a hyperbola.
6. As the gas is compressed, the temperature should rise so the volume will be larger than what has been assumed here. However, the cylinder is small and the addition of force occurs slowly so the temperature stays relatively constant. Some students might have noticed that when the masses are removed quickly, the piston does not return to its original position until later.

LAB QUIZ ANSWERS

1. c
2. c
3. d
4. b
5. a

Lab #35

Name: ______________________

Date: ______________________

Using an Echo to Find the Speed of Sound

In this lab you will use an echo to investigate the speed of sound. By timing how long it takes for sound to return to you, you can find the speed of sound. Because of problems with timing, your results will not be very precise but will give a good approximation of the speed. Using an echo doubles the distance the sound travels. It also allows the timer to be near the source of sound (your lab partner making a sound) and far from it at the same time.

Materials

place for a good echo

meter stick

timer which measures to tenths of a second

Procedure

1. Find a place with a good echo where you can hear the sound returning to you from at least a distance of 150 m. A building with a large flat side provides a nice reflective surface for sound waves. Better yet is a building which is not rectangular, but has a wing. This means that there is an inside corner to the building as shown in Figure 35.
2. Make sure that you are not going to disturb other people with loud noise.
3. Measure the distance which you walk in ten paces. Record this value.
4. Walk as far from the building wall or the inside corner of the building as you can and still hear an echo. Count the number of paces which you walk and try to keep them as similar in size as possible. Record this number.
5. Have your partner make a noise. When you hear the noise, start your timer. When you hear the echo, stop the timer. Repeat this a number of times and record each time.
6. You should try to start and stop the timer in the same manner each time. This means that you should not anticipate when your partner is going to make the noise. You should hear the noise and then start the timer. Likewise, you should hear the echo and then stop the timer. In this way, you start the timer late because of your reaction time but you also stop the timer late because of reaction time. (The reaction time is the time which it takes to hear the noise and start or stop the clock.)

(continued)

Figure 35

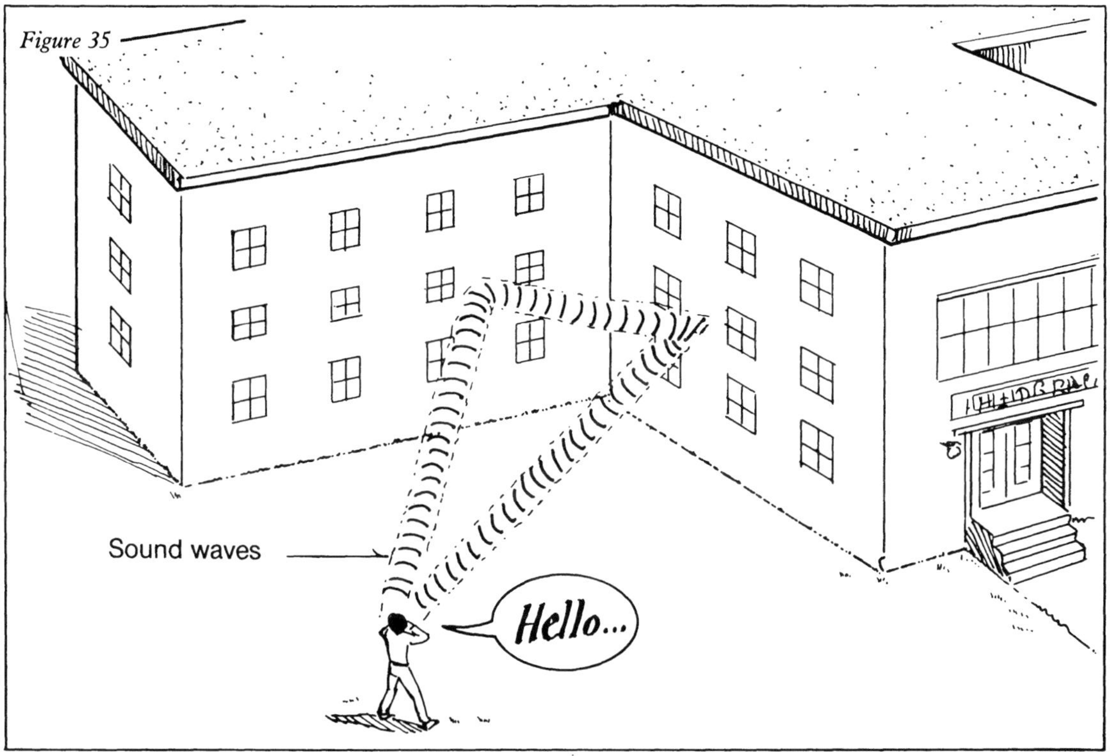

Caution

Don't disturb other people as you make noise.

Analysis

1. Use the distance for ten paces to calculate the distance from the building to where you made your noise.
2. Because sound travels to the building and back, multiply this value by two to find the distance which the sound travels.
3. Find the average time for the sound to return to you.
4. Find the speed of sound. Make sure that you use the proper number of significant digits. Remember that your measurements are not very precise.

(continued)

Lab #35
Using an Echo to Find the Speed of Sound *(continued)*

Name: ______________________

Date: ______________________

Conclusions

1. Do you think that your value for the speed of sound is reasonable, given the way that you made your measurements? Explain.
2. How could you improve the accuracy of your answer by making simple changes in the experiment? Explain.

Quiz #35

Name: ______________________

Date: ______________________

Using an Echo to Find the Speed of Sound

1. An echo is used to

   a. double the distance the sound travels.
   b. triple the distance the sound travels.
   c. make sound travel faster.
   d. make sound travel slower.

2. You measure distances in this lab with a meter stick and

   a. a tape measure.
   b. a laser.
   c. paces.
   d. a wheel.

3. The time between when you hear a noise and start or stop the timer is called the ______________ in this lab.

   a. starting time
   b. stopping time
   c. noise time
   d. reaction time

4. You must run the lab in such a way that the noises which you make

   a. do not travel too fast.
   b. do not travel too slow.
   c. do not disturb anyone.
   d. are of the right frequency to make an echo.

5. In a particular medium, sound traveled 800 m in .5 sec. The velocity of the sound in the medium in m/sec was

   a. 400.
   b. 800.
   c. 1200.
   d. 1600.

Lab #35: Using an Echo to Find the Speed of Sound

Teacher Notes

> ***Caution***
>
> Don't disturb other people as you make noise.

This lab is fun if there is an appropriate place to do it. It is a useful lab even though it doesn't produce very accurate results and students can actually make up their data if they do it at home. The timing must be done to the nearest tenth of a second. Check to see how many students have inexpensive watches with a stopwatch function. A good supply of these watches will allow students to make the measurements on their own.

You may want your students to do the lab at home if they have appropriate surroundings. For those living in the city, you may need to modify the experiment because other noise may drown out the echo. If, for instance, a student can stand on one end of a block and listen for a noise from a bus which comes into view, then there is an event and a noise associated with the event. (Caution students to do the lab in a safe place.) The experiment does not depend on an echo, but the distance may be large enough to get good data. The only real requirement is that the time of travel for the sound is at least a second. Therefore, the sound must travel for at least 300 m.

Because students should not disturb other classes with loud noises, doing the lab may be difficult at school. Making a noise by slapping two boards together might be more acceptable than a student voice.

Possible Problems

1. If the distance is not great, then the time is too small.
2. The lab must be conducted so that others are not disturbed.
3. Students may make up data instead of actually getting them. For this lab, making up data is relatively easy and is hard to spot.

Conclusions

1. Yes, the errors may be as much as 20–30% and still be reasonable, given the timing error. The error in distance should be no more than 5%.

(continued)

Teacher Notes (continued)

2. You could use a longer distance or you could use some kind of electronic timing to get a more precise value for time. Both of these would reduce the percentage error in timing.

LAB QUIZ ANSWERS

1. a
2. c
3. d
4. c
5. d

Lab #36
Waves in Strings

Name: ____________________

Date: ____________________

In this lab you will investigate the formation of waves in a string. You will use a buzzer to cause a string to vibrate and by changing the tension in the string you will cause different standing waves to form. The different tension will cause the wave to travel at a different speed so different numbers of half waves can exist in a given length of string.

Materials

clamp	string
buzzer	power source
small stick of wood	masses

Procedure

1. Clamp your buzzer about two thirds of a meter from an edge of the lab bench. Place the small stick at the edge of the table (see Figure 36).
2. Tie a 1-m-long string to the buzzer. Then tie a small loop in it at the other end and let the loop drop over the piece of wood.
3. Turn the power supply on and then pull the string gently, slowly increasing the force. You should pull the string so that it goes from the buzzer over the piece of wood and then down to your hand. As you pull the string, watch for the formation of nodes and antinodes. They form and then disappear as the force changes. Pull the string until you have two antinodes in it.
4. In a string with fixed ends, the number of antinodes equals the harmonic—the second harmonic has two antinodes, the third has three, etc. With some buzzers it is possible to get just one antinode (the first harmonic); most, however, will not buzz when the force is large enough to make one antinode. If you pull too hard, you stop the buzzer or break the string. Reduce the force if the buzzer stops.
5. After you have experimented with making the different numbers of nodes and antinodes, get values for the force which creates each standing wave. Do this by adding mass to the loop hanging down from the edge of the bench. The mass exerts a force equal to its gravitational force. Adding the mass takes a bit of patience because the standing waves have the tendency to come and go. Start by finding the mass required to cause the second harmonic. When you have found the mass, record it and the harmonic.

(continued)

Lab #36
Waves in Strings *(continued)*

Name: ______________________

Date: ______________________

Figure 36

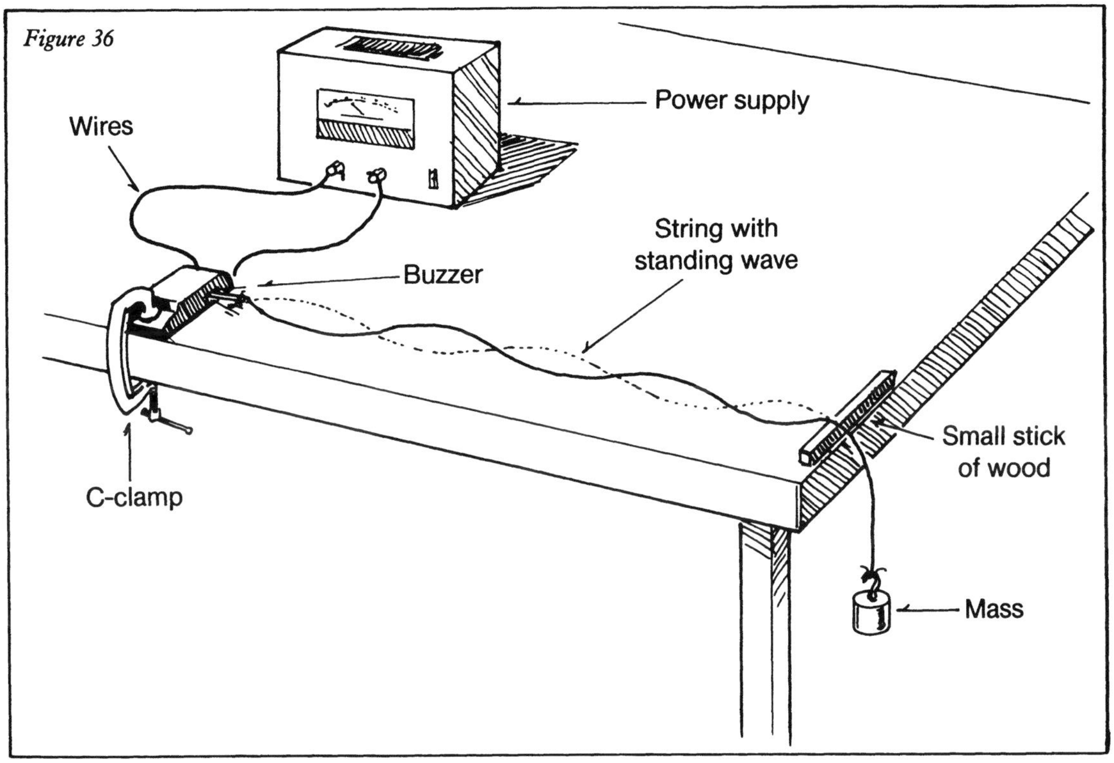

6. Next find the mass required to make the third harmonic. Record the mass and harmonic number. Then find the mass needed on the end of the string for as many other harmonics as possible. Some of the harmonics are multiples of other harmonics, like 6 which is twice 3. Multiples of harmonics are often hard to form. For some of your measurements, you can measure to smaller units of mass than you actually put on the loop. If the standing wave forms just after you start slowly placing a weight on the string and then disappears when you have let go of the mass, then the mass required to cause the standing wave is a fraction of the mass you added.
7. Remove the masses and then use your hand to exert a force so that the second harmonic forms. Place your finger at the node in the middle of the string. Describe what happens.
8. Finally, start with no force on the string and then slowly pull the string until the second harmonic forms. Describe what happens to the string as the force is increased. Make particular note of the way the string behaves near the buzzer.

(continued)

Lab #36

Waves in Strings *(continued)*

Name: ______________________

Date: ______________________

Caution

Don't add so much mass that you break the string. Your teacher can tell you the maximum value.

Analysis

1. For each of the masses which you used, find the gravitational force.
2. Plot a graph of force vs. the number of harmonics with the number of harmonics on the horizontal axis.
3. Find a value for the square root of each of these forces.
4. Find a value for the inverse of the square root of the force ($1/\sqrt{F}$).
5. Plot a graph of the inverse of the square root vs. the number of harmonics with the number of harmonics on the horizontal axis.

Conclusions

1. How does the force on the string change as the number of harmonics changes?
2. Using your graph, predict the force which would be needed to produce the first harmonic. Explain how you got your answer.
3. The frequency of the buzzer stays almost constant. The equation for the frequency of waves in strings is:

   $$f = (n/2L)\sqrt{F/\mu}$$

   where f is the frequency, n is the harmonic number, L is the length of the string, F is the force in the string, and μ is the mass per unit length of the string. For this experiment, frequency, length, and mass per unit length (f, L, and μ) are constant. This means that the only two variables are the number of harmonics and the force. These are related with n times square root of F equaling a constant: $n\sqrt{F} = k$. Therefore, the graph of the inverse of the square root of the force vs. the number of antinodes should be a straight line. Does your graph show this? Explain.
4. What happens when you place your finger at a node? What changes take place?

(continued)

Lab #36

Waves in Strings *(continued)*

Name: ______________________

Date: ______________________

5. Describe what happens in the string as the tension in the string is increased.
6. Is the place where the string is attached to the buzzer a node or an antinode? Explain.

Quiz #36
Waves in Strings

Name: ______________________________

Date: ______________________________

1. In this lab, waves in a string are created by
   a. you as you move your hand rapidly.
   b. a rotating platform.
   c. a changing electric field.
   d. a buzzer.

2. In this lab, different forces on the string will produce
   a. different numbers of antinodes.
   b. the same number of antinodes.
   c. a large change in the frequency.
   d. a large change in the length of the string.

3. If there are four antinodes in the string, you must be observing the ____________________ harmonic.
   a. first
   b. second
   c. third
   d. fourth

4. Exerting the proper amount of force to get a harmonic in this lab is
   a. easy.
   b. a bit difficult.
   c. sometimes quite difficult.
   d. impossible.

5. In this lab, you will
   a. see different harmonics form in the string.
   b. see different numbers of antinodes form in the string.
   c. exert different forces on the string.
   d. do all of the above.

Lab #36: Waves in Strings

Teacher Notes

Caution

Don't add so much mass that you break the string. Your teacher can tell you the maximum value.

In this lab students learn how changing the force on a string results in a different harmonic forming in the string. It is not an easy lab to do because students may be frustrated taking data when the forces are small. However, students do enjoy seeing the change in harmonics as they exert different forces, especially because they seem to jump up out of the string when the force is right.

Before you attempt to do this lab, make sure that you are comfortable with it and that you can get the proper results with your equipment. Most buzzers which can be purchased for under $10 from supply houses will work. A much heavier-duty buzzer is available from Central Scientific (called a "string vibrator," #85102). It maintains a more constant frequency than the less expensive buzzers and, since it is more energetic, it may be possible to find the first harmonic. If the lab is conducted as a demonstration rather than as a group data-gathering activity, the $80 investment may be worth it.

You must have some masses which are small (about a gram). You may want to use paper clips which are less than a gram, or washers. Perhaps a combination is best. The mass of the paper clips and/or washers can be found by finding the mass of a large number of them. You need 10 gram masses plus larger ones. Make sure that the buzzers can be easily clamped in place and that they are strong enough to vibrate the string at the second harmonic. The small block of wood placed at the edge of the table raises the string off the table and allows the string to vibrate more easily at that end. It needs to be about the size of a pencil.

When you experiment with this setup, you will probably find some interesting facts about how the antinodes form. It is difficult to form harmonics which are multiples of others. With one particular buzzer which cost about $3, it is possible to get the fourth, fifth, seventh, and ninth harmonics, but not the sixth, eighth, or tenth. With some buzzers, it is possible to get all of the harmonics, but with inexpensive ones the string has a tendency to drive the buzzer. Instead of getting the eighth, you get the fourth; you can almost see the eighth form before it disappears.

Because of the tricky nature of the data gathering, you may wish to conduct this lab as a demonstration from which all students take the same data. This will allow you

(continued)

to point out some of the interesting points to all students. If you have sensitive spring scales, you can also use them to find the force. Doing the lab as a demonstration allows you to do some extra things such as using a strobe to find the frequency. (Increase the strobe frequency until you see the string as a crest and trough. When this happens, you are at twice the frequency of the string. Reduce the strobe frequency by a factor of two and you should see just one image. This is the frequency of vibration.)

Possible Problems

1. Students may have difficulty measuring the forces. Keep encouraging them.
2. The buzzer may not keep buzzing when you try to find the force for the second harmonic. You may be able to increase the buzzer voltage or lengthen the string.

Sample data: (string length .6 m, inexpensive buzzer, cotton string)

Harmonic:	2nd	3rd	4th	5th	7th	9th
Force (N):	.22	.12	.080	.049	.023	.015

Conclusions

1. The force becomes less as the number of antinodes increases.
2. The force for the first harmonic should be about four times the force for two antinodes. This can be found by extrapolating the graph or using the equation.
3. Yes, the line is a straight line which agrees with the equation.
4. The part of the string near the buzzer may keep buzzing but the other half of the string stops. The string may continue to vibrate unchanged, or it may ruin the standing wave. Students should have a combination of two of these if they are good observers.
5. The standing waves form and disappear in an almost ghostlike way.
6. The motion of the buzzer causes that end of the string to be more like an antinode than a node. A node exists at the edge of the table.

(continued)

Teacher Notes (continued)

LAB QUIZ ANSWERS

1. d
2. a
3. d
4. c
5. d

Lab #37
Resonance in Pipes

Name: ______________________________

Date: ______________________________

In this lab you will investigate how pipes resonate. You will hold a tuning fork over a pipe held vertically in a bucket of water (see Figure 37). By moving the pipe up and down, you can change the length of the column of air in the pipe. With some lengths, a standing wave forms the same frequency as the tuning fork. This standing wave will amplify the sound of the tuning fork so sound can be heard. By measuring the length of the pipe, you can calculate the speed of sound in the pipe.

Materials	
bucket	tuning forks
water	meter stick
various lengths of plastic pipe	

Procedure

1. Fill the bucket with water to near the top. Then take the smallest piece of plastic pipe and put it in the water so it touches the bottom.
2. Strike the tuning fork on your palm and hold it about 1 cm above the pipe. Move the tuning fork and pipe upward keeping the tuning fork about 1 cm above the pipe. You will probably have to strike the tuning fork on you palm every 10 seconds or so in order to keep it vibrating. Listen for a louder sound coming from the pipe. If you hear one, then find the position where the sound is loudest. If you do not hear a loud sound, then take the next longer piece of pipe, put it to the bottom of the bucket, and then start pulling it out.
3. When you have made a loud sound, you have found the first harmonic. Because there is an out-of-phase reflection at the water, the pipe contains one quarter of a wavelength. Measure the distance from the tuning fork to the water for the first harmonic. Record this distance and the frequency of the tuning fork.
4. For a closed-ended pipe, the next harmonic is the third, which forms in a pipe three times the length of one for the first harmonic. Select a piece of pipe and find the third harmonic. When you find it, record the actual distance from the top of the pipe to the water, the frequency, and that this is for the third harmonic. If you have a long-enough pipe, try to find the fifth harmonic, which requires a pipe five times the length of one for the first harmonic.

   Note: If you can't find the third harmonic, it may be that what you thought was the first harmonic was actually the third harmonic. If you don't find a loud noise

(continued)

Lab #37
Resonance in Pipes *(continued)*

Name: ______________________

Date: ______________________

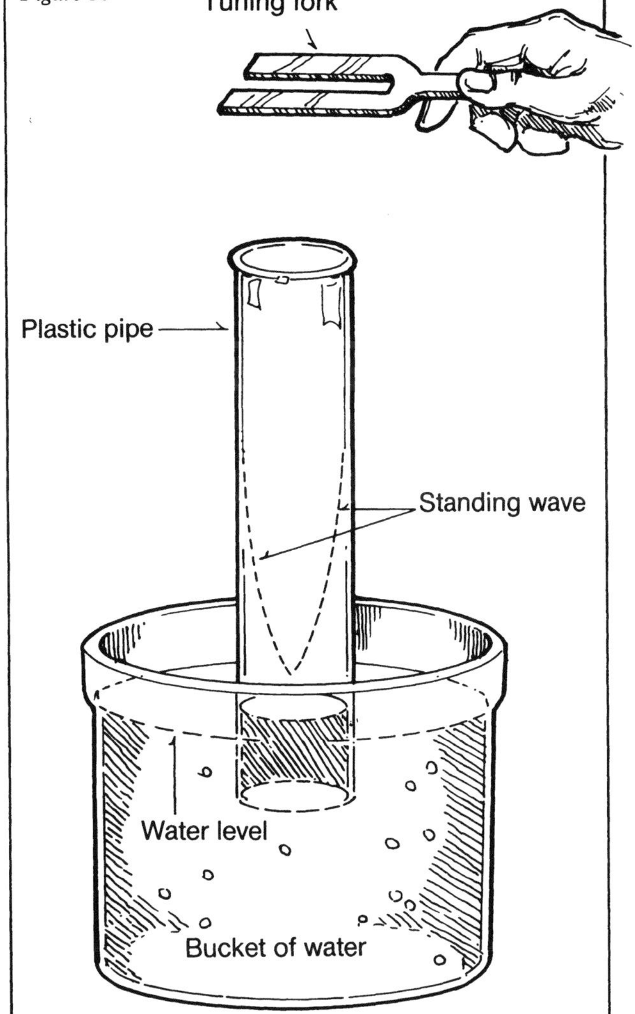

Figure 37

when the length is three times longer, try dividing the length you found in step 3 by three.

5. Repeat steps 3 and 4 for a second tuning fork.
6. Repeat steps 3 and 4 for a third tuning fork which has an unknown frequency.

(continued)

Name: ______________________

Date: ______________________

Caution

Do not strike the tuning forks on anything other than your palm. Striking them on sharp objects can break them.

Analysis

1. Using the data from the first tuning fork, find the speed of sound in your pipe using the equation: $v = f\lambda$.
2. Using the speed of sound which you found in step 1 and the length of the pipe which gave you the first harmonic for the tuning fork of unknown frequency, find the frequency of the tuning fork.

Conclusions

1. How well does the speed of sound agree with the normal value of about 330 m/sec at room temperature? Explain any differences.
2. What is the frequency of the tuning fork with the unknown frequency? Indicate any possible sources of error.
3. Were the lengths for the third harmonics actually three times as large as the first harmonic? Explain the differences.

Quiz #37

Resonance in Pipes

Name: ______________________________

Date: ______________________________

1. The length of a column of air in a pipe is changed by

   a. moving your hand in and out.
   b. placing a small disk in the pipe.
   c. cutting the pipe.
   d. moving the pipe up and down in water.

2. You know when you have found one of the harmonics because

   a. it is marked on the pipe.
   b. it is observed by the motion of the water.
   c. the small disk moves more violently.
   d. you hear a louder sound.

3. If a pipe has a quarter of a wavelength in it and the pipe is 2 m long, what is the wavelength of the wave in meters?

   a. .5
   b. 1
   c. 4
   d. 8

4. A wave has a frequency of 100 hz and a wavelength of 2 m. What is the wave speed in m/sec?

   a. 50
   b. 200
   c. 400
   d. 800

5. There is danger of breaking apparatus in this lab. What is the apparatus you are most likely to break if you do not follow the procedure?

   a. the pipe
   b. the timer
   c. the tuning fork
   d. the water container

Lab #37: Resonance in Pipes

Teacher Notes

Caution

Do not strike the tuning forks on anything other than your palm. Striking them on sharp objects can break them.

In this lab students work with standing waves and actually hear them. This is a much better approach than a demonstration of the effects, and the data are generally good. One problem with the lab, however, is that if you have eight or ten sets of the lab materials, you may have a storage problem because the pipes take up some room. Except for this limitation, the lab is easy to run and relatively easy to set up, and the materials are not expensive.

You need a variety of tuning forks, which most physics departments have in good supply. Science Kits supplies them for about $8, but you may be able to get a better price through a music store. The water containers can be anything which will hold water.

The pipe, which is easy to use and is relatively inexpensive, is 2-inch black PVC pipe used for plumbing. It comes in 10- or 20-foot lengths and can easily be cut with a saw. (A regular carpenter saw works well.) You need to have cut pieces so that students can get any length of pipe that they need for the container of water which you use. The deeper the container, the fewer the pieces required. The first piece should be about two inches longer than the depth of the water container; the next piece should be twice the water depth; the third should be three times the depth, etc. The actual length and size will depend on the size of your water container. Do a quick calculation using the frequency of the lowest-frequency tuning fork which you want to use to find the length of the tube for the third harmonic, which is three quarters of a wave.

The "unknown" frequency tuning fork can be made using tape to cover the actual frequency.

Possible Problems

1. Students may not be able to hear the standing wave. You may have to insist on keeping the noise level down in the lab.
2. If students hit the tuning forks on something other than their palm (which has a direct feedback loop so they don't hit it too hard), the tuning forks may break.

(continued)

Conclusions

1. The speed of sound should agree quite well with the accepted value.
2. The unknown frequency can be estimated to within 10% without much trouble.
3. These lengths should be close to the theoretical values. There is a problem because the effective length of the tube is not exactly the length of the tube. This is associated with frictional effects along the sides of the tube.

LAB QUIZ ANSWERS

1. d
2. d
3. d
4. b
5. c

Lab #38

Reflection and Refraction: Tracing Light Rays

Name: ______________________

Date: ______________________

In this lab you will investigate how rays of light reflect off a mirror, and how they travel as they enter a new medium and then exit from it. Rays are traced using pins to follow the path of the light. You will measure the angles of incidence, reflection, or refraction. You will also calculate the index of refraction of the liquid and the velocity of light in water, given the velocity of light in air is 3×10^8 m/sec.

Materials

cardboard	pencil
paper	protractor
mirror	plastic container
pins	water

Procedure

Case 1: Reflection

1. Place a piece of cardboard under a piece of paper and stand the mirror up using pins or other devices so that it is vertical as in Figure 38.1.

Figure 38.1

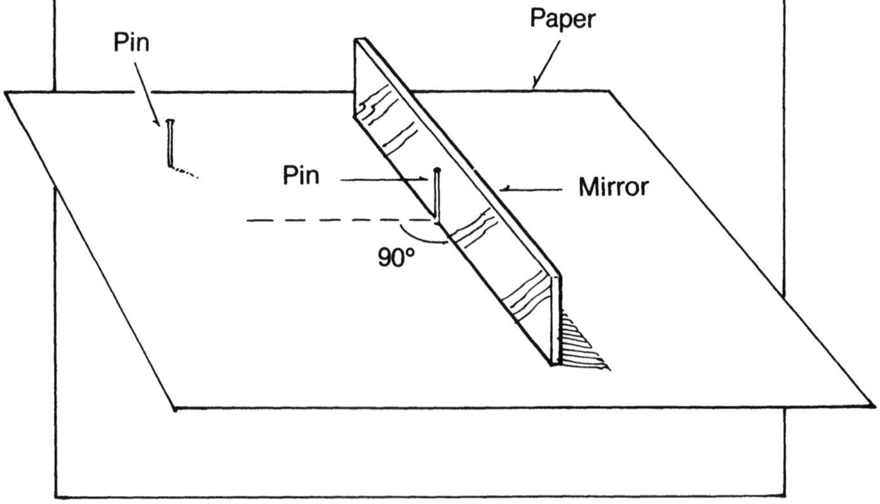

(continued)

Name: ______________________________

Date: ______________________________

2. Mark the position of the mirror by drawing a line along the back side of the mirror. This is done because the back side of the mirror is the surface which actually reflects the light.
3. Place a pin next to the mirror on the reflecting side and draw an imaginary line perpendicular to the mirror which goes through the pin. Then place another pin about 30 degrees away from this perpendicular about 5 cm from the mirror.
4. Look into the mirror and find the two pins. Place a third pin about 5 cm from the mirror so that it is in line with the reflection of both of the other pins.
5. Remove the mirror and draw a line which goes through the holes from the first two pins and goes to the line which represents the back of the mirror. Then draw a line from where this first line intersects the mirror to the hole from the third pin. This traces the path of light from the first pin to the mirror, and back toward your eye.
6. Draw a line perpendicular to the line which represents the mirror where the light rays strike it. Then find the angle of incidence and the angle of reflection.

Case 2: Refraction

7. Draw an outline of the container in the shape of a half circle on a piece of paper.
8. Remove the container and with a protractor, draw a line which is perpendicular to the container's surface where you have placed your pin. This line should be outside and inside of the outline of the container which you drew. Now replace the container, being careful to return it to its original position. Then add water to the container so that it is almost full.
9. Place a pin at the center of the flat side. Then place a second pin about 5 cm from the first pin, but about 30 degrees away from the perpendicular which you drew as in Figure 38.2. Then look through the container and place a third pin next to the curved part of the container so that it is in a line with the others. Make sure that you are looking *through* the water and not over the top.
10. Place a fourth pin 5 cm from the container so that it is in line with the other three pins. All four pins should appear to be in a straight line when you sight them with your eye and look through the liquid. When you do not look through the water, the pins will not appear to be in a straight line.
11. Remove the container and connect the dots made by the four pins. They will make a path which the light took from the first pin, through the water, to the last pin.

(continued)

Name: ______________________

Date: ______________________

12. Measure all angles of incidence and refraction for the light as it goes into the container of water and then as it goes out. The angle at which it leaves the container should be zero.

Figure 38.2

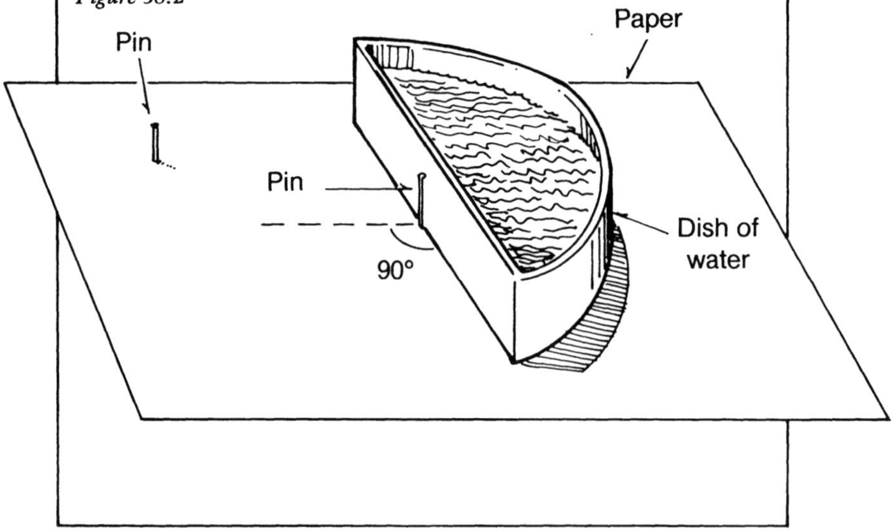

Case 3: A Lens

13. Place the container on a piece of paper and draw a line around it. Then draw three lines perpendicular to the flat side, one going through the center as shown in Figure 38.3. The other two lines should be about 3 cm away from the center, one on one side of the center and one on the other. Place pins where each of these perpendicular lines meet the container. Then place three pins about 3 cm from the container on the lines you just drew.
14. Add water to the container and look through it at a pair of pins on one of the lines. Place a pin in line with two pins on your side of the container right next to it. Then place another pin about 6 cm from the container so that all four pins are in a line as viewed through the water. Repeat this for the other two pairs of pins.
15. Remove the container and connect the pin holes with lines.

(continued)

Name: ______________________

Date: ______________________

Figure 38.3

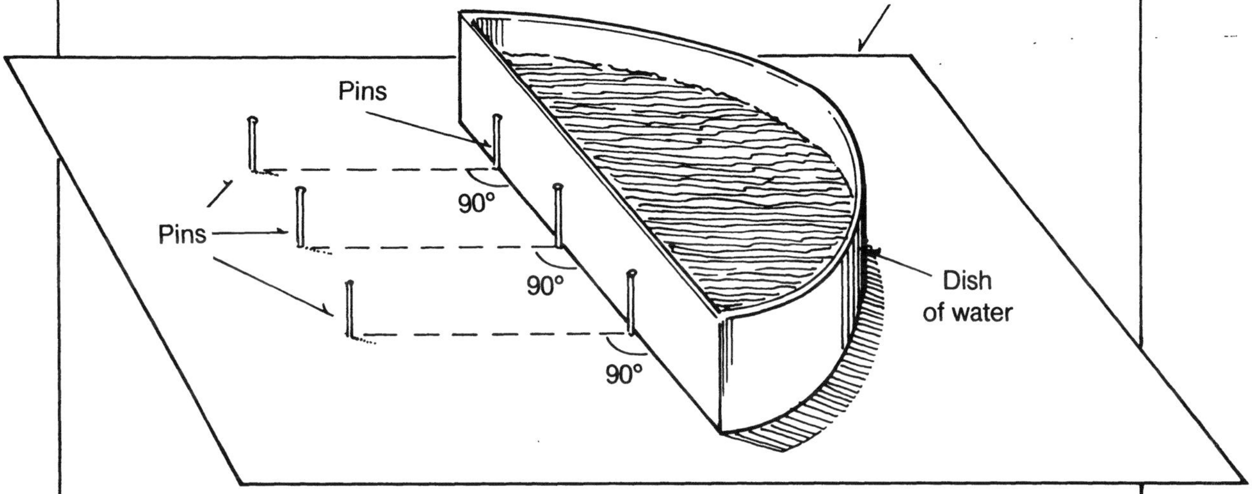

Caution

Be careful when placing your pins and ask any questions about the procedure if you are unsure. Measure all angles with the protractor.

Analysis

Using the equation: $N = \sin(\theta_{air})/\sin(\theta_{medium})$, find the index of refraction for water using the angles you found in Case 2.

(continued)

Lab #38

Reflection and Refraction: Tracing Light Rays *(continued)*

Name: ______________________

Date: ______________________

Conclusions

1. In Case 1, does the angle of incidence equal the angle of reflection? Give data to support your conclusion.
2. In Case 2, describe what happened to the light as it went through the water.
3. What error do you estimate for your speed of light in water? Explain.
4. In Case 3, what happens to the three parallel rays of light as they go through the container of water? How is this like a lens?

Quiz #38

Reflection and Refraction: Tracing Light Rays

Name: ______________________

Date: ______________________

1. When light waves reflect, the angle of incidence is ______________ the angle of reflection.

   a. larger than
   b. smaller than
   c. equal to
   d. sometimes larger than

2. When light waves refract and go into a slower medium, the angle of incidence is ______________ the angle of refraction.

   a. larger than
   b. smaller than
   c. equal to
   d. sometimes larger than

3. In this lab, you use ______________ to trace rays of light.

   a. a laser
   b. a photometer
   c. a protractor
   d. pins

4. The reason why light waves refract is that

   a. the velocity of light changes.
   b. the frequency of the light changes.
   c. when light hits a surface it reflects.
   d. they enter a substance which does not allow light to travel.

5. A place where light rays converge is called a

   a. point.
   b. focus.
   c. convergence.
   d. origin.

Lab #38: Relfection and Refraction: Tracing Light Rays

Teacher Notes

Cautions

Be careful when placing your pins and ask any questions about the procedure if you are unsure. Measure all angles with the protractor.

In this lab students follow rays of light as they reflect and refract. Many times they see what is happening for the first time, especially with the lens section (Case 3). It may take them a little while to figure out exactly how to place the pins, but once they get the idea, they can trace the rays in their mind quite a bit better.

The equipment is not complicated—small mirrors about four or five inches long and an inch high work well. It is helpful if you have some which are mounted so the students do not have to support them with pins. A plastic dish which is a half circle is ideal. Central Scientific has these dishes, called semicircular cells, for about $4. A semicircular lens can also serve the same function but is more expensive (about $12). It is possible to examine the reflection off a curved mirror but it seems to be a little more difficult for students to understand.

Possible Problems

1. Students must be reminded to measure the angles to the perpendicular and not to the surface.
2. Because most inexpensive mirrors are "rear surfaced," students must take care as they trace the path of the reflecting rays of light. Placing a pin at the back of the mirror may help some to note the actual place of reflection to draw the diagram properly.
3. Students may tend to look at the pins over the water instead of through it.
4. Students have been told to place pins at "about" 30 degrees from the perpendicular. They may do this but fail to find the actual angle with the protractor.

Conclusions

1. The angle of incidence should equal the angle of reflection within a degree or so. It may be off because the student did not realize that it is a back-surfaced mirror.
2. The ray is refracted as it changes medium.

(continued)

Teacher Notes (continued)

3. A 10% error in the speed of light comes from the measurement of angle.
4. The parallel rays of light come together at a focus which is like the way rays of light meet at a focus for a lens.

LAB QUIZ ANSWERS
1. c
2. b
3. d
4. a
5. b

Lab #39

Lenses

Name: ______________________

Date: ______________________

In this lab you will investigate lenses and how images are formed by them. As shown in Figure 39 you will use a lens, a light source, an object, and a screen to observe images from lenses and to check on the validity of the equation:

$$l/f = l/s_o = l/s_i$$

You will first find the focal length of your lens and then you will use this information to find the approximate positions of the images that you are looking for.

Figure 39

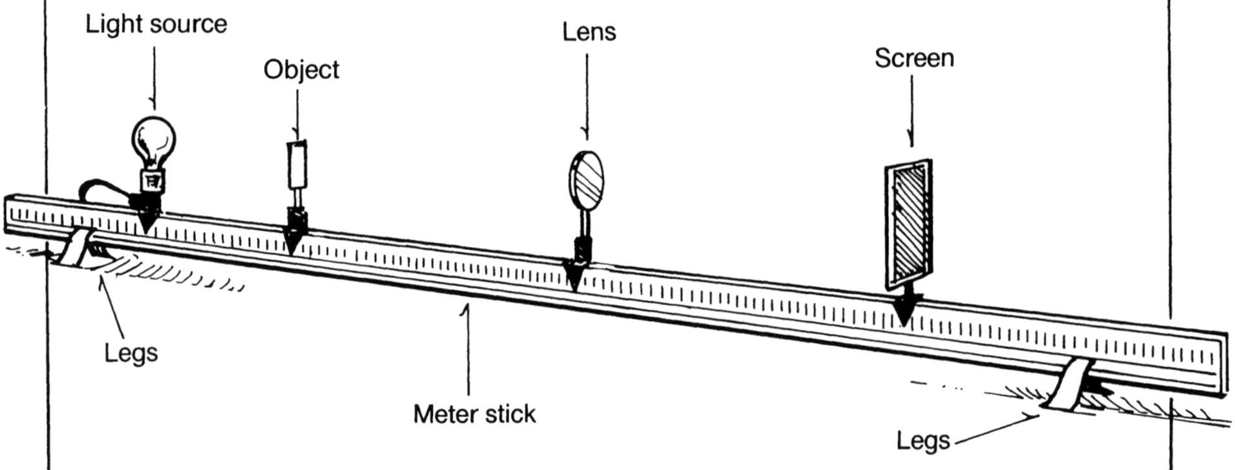

Materials

- lens
- optical bench
- light source
- object
- lens holder
- white screen

Procedure

1. Take your lens and select a bright object that is a long distance away. Project an image of this object on a flat surface such as a wall, getting as sharp an image as

(continued)

Name: ______________________________

Date: ______________________________

possible. Measure the distance from the lens to the wall precisely. Record this value.

Case 1:

2. Set up the optical bench so that light from the source hits the object, and then goes approximately twice the focal length to the lens, which is held in a lens holder.
3. Turn on the light source and then, using the white screen, find the position of the image (it should be about twice the focal length on the opposite side of the lens from the object). Precisely measure and then record the object and image distances. Note the type of image you see (enlarged or reduced, real or virtual, and inverted or upright).
4. With your hand, cover up half of the lens. Record your observations of what happens to the image on the screen.

Case 2:

5. Move the lens so that it is about 3 focal lengths from the object. Find the position of the image with the screen. (It should be between f and $2f$.) Precisely measure and then record the object and image distances. Note the type of image you see (enlarged or reduced, real or virtual, and inverted or upright).

Case 3:

6. Move the lens so that it is about 1.5 focal lengths from the object and then find the position of the image with the screen. Precisely measure and then record the object and image distances. Note the type of image you see (enlarged or reduced, real or virtual, and inverted or upright).

Case 4:

7. Move the lens so that it is about half a focal length from the object. A real image will not be formed with this setup. You will not be able to find the position with the screen. The position must be found by other means.
8. Look through the lens at the object. You will see the image of the object. (Make sure the rays of light that arrive at your eye go through the lens.) Have your partner hold a pencil or other pointed object just to the side of the lens. You will see the image of the object through the lens and the pencil beside the lens. The image should appear farther away from you than the pencil.
9. Have your partner move the pencil slowly away from you. At some point, you will see that the pencil (which is not viewed through the lens) and the image of the object (which is viewed through the lens) will be at the same distance from

(continued)

you. (This requires that you compare the distances that the image and the pencil are away from you. This may be hard to see at first, but once you see it, it will be quite obvious.) Record this position, the distance from the lens to the "image" of the object. (This distance should be approximately 1 focal length.) Note the type of image you see (enlarged or reduced, real or virtual, and inverted or upright).

10. Repeat the steps for Case 4 so that your partner can see the position of the image and pencil.

Caution

Do not drop the lenses because they will break.

Analysis

For each case, use the lens equation to find the theoretical image distance.

Conclusions

1. Do the image distances you found by experiment agree with those predicted by the equation? Explain any differences which exist.
2. Do the data seem to support the equation?
3. Describe in your own words the way that you saw the image of step 9. How did you know that the image and your partner's pencil were in the same place?

Quiz #39
Lenses

Name: ______________________

Date: ______________________

1. Parallel rays of light going through a converging lens
   a. diverge.
   b. meet at a focus.
   c. continue parallel.
   d. form an image.

2. If the image from a lens is not real, it must be
   a. virtual.
   b. upright.
   c. enlarged.
   d. focal.

3. If the image from a lens is not inverted, it must be
   a. virtual.
   b. upright.
   c. enlarged.
   d. focal.

4. The only measurements which you will make in this lab are measurements of
   a. optical quality.
   b. focusing ability.
   c. optical intensity.
   d. distance.

5. Which image may be hard to see at first?
   a. the inverted image
   b. the enlarged image
   c. the virtual image
   d. the composed image

Lab #39: Lenses

Teacher Notes

Caution

Do not drop the lenses because they will break.

This lab is a good introduction to the images produced by a converging lens. It is rather easy to set up and the results are generally good. Students do have a hard time finding the virtual image distance, but this might be expected because they have never had experience with this kind of measurement. They have used magnifying glasses before, but they probably never have thought about where the image really is.

The best inexpensive lab bench uses a meter stick for the optical bench. The supports, lenses, lens holders, light source, and object screen can all be purchased from physics supply companies for about $30. For $10 less, you can get setups which use candles for light sources; however, candles are more of a problem to use and are more hazardous. The 7.5-cm lenses give brighter images and are easier for students to use but the 3.75-cm lenses are usable. The focal length for the lenses should be from 15 cm to 20 cm to work well on the meter stick. Most physics labs do not need to buy the complete outfit because it includes meter sticks as well as the other materials.

The virtual image in Case 4 is not easy for most students to see. Make sure that you have experimented with seeing this image before the lab. You must be able to tell them exactly what they should be looking for. It may help to show them a magnifying glass and how the image seems farther from them than it really is. After they have seen one of these images and where it is, then they can measure the virtual image distance more easily.

Possible Problems

1. Students may not precisely measure their focal length which is the basis for all their calculations.
2. Students may not know what the object is and measure to the light bulb in the light source rather than the object.
3. Students may not look at the object through the lens when they are trying to see the position of the virtual image.

(continued)

Teacher Notes (continued)

Conclusions

1. The equations should predict the actual values with less than 5 or 10% error. Measurement error, especially of the focal length, is generally the problem.
2. The data should support the equation.
3. The description of how they see the virtual image should be similar to the one given in steps 8 and 9.

LAB QUIZ ANSWERS

1. b
2. a
3. b
4. d
5. c

Lab #40
Telescopes and Microscopes

Name: ______________________________

Date: ______________________________

In this lab you will investigate how telescopes and microscopes are made. You will use an optical bench to build these devices using two lenses—the primary and the secondary, or eyepiece. Although the images you will get from them will not be perfect, they will be as good as the telescope Galileo used when he looked at the moon, Jupiter, and the heavens.

Materials	
lenses	screen
lens holders	light source
optical bench	white card

Procedure

1. Take each lens and project an image of a distant object on a wall. Get as sharp an image as you can and then measure the distance from the lens to the wall. Record this value. The distance is the focal length. Then put each lens in a lens holder.

Case 1: Microscope

2. Set up the optical bench and place an object in front of the light source (see Figure 40.1).

Figure 40.1

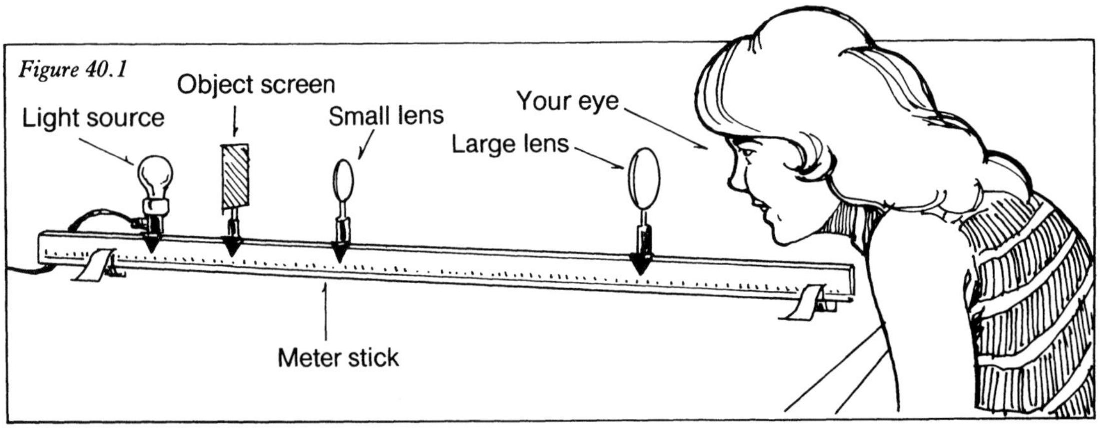

(continued)

Name: ______________________

Date: ______________________

3. Place the lens with the small focal length about one centimeter farther from the object than its focal length. This lens is called the primary. Use a white card to find where the image of the object is. Record this value. Describe the image.
4. Remove the white card and then take the lens with the large focal length, called the secondary, and place it where the image from the first lens was formed. Look through the large lens and toward the object. What do you see? Then very slowly move the secondary lens away from the primary lens. Write a description of what you see. Move it so that the large lens is more than one focal length from the place where the primary image was formed. Note the positions of significant events. (An event would be when you can or can't see an image, etc.)
5. Move the primary lens .5 cm closer to the object. Look at the object through the lenses. Move the secondary lens until you get the best image. What do you observe? Experiment further with moving the primary lens and describe the results. What combination of positions of the primary and secondary give the most magnification?

Case 2: Telescope

6. Move your optical bench so that one end of it points toward a window (see Figure 40.2). The things that you see outside the window will be your object. Remove

Figure 40.2

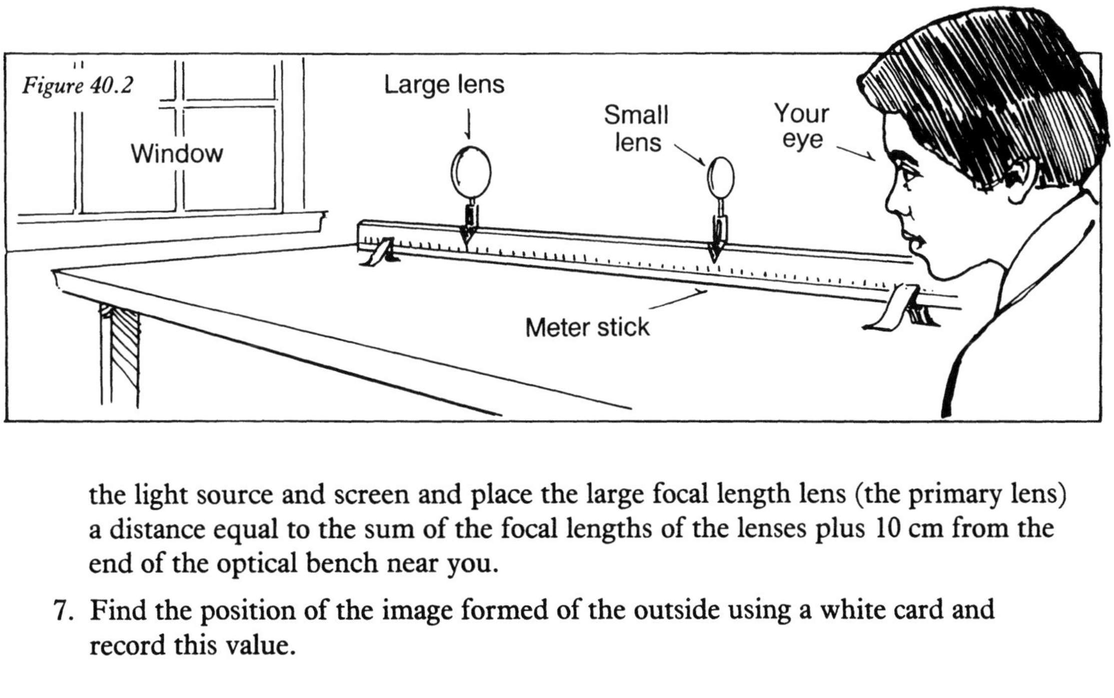

the light source and screen and place the large focal length lens (the primary lens) a distance equal to the sum of the focal lengths of the lenses plus 10 cm from the end of the optical bench near you.

7. Find the position of the image formed of the outside using a white card and record this value.

(continued)

Lab #40

Name: ______________________

Telescopes and Microscopes *(continued)*

Date: ______________________

8. Place the small focal length lens (the secondary lens) where the primary lens focuses the image of the outside. Look through the secondary lens toward the primary lens and then slowly move the secondary lens away from the primary. Look through the lens and describe what you see as you move it farther and farther away.
9. Try different focal length secondary lenses. Describe the results.

Cautions

1. Not being careful in your measurement of the focal lengths will cause problems.
2. Do not drop the lenses because they will break.

Analysis

1. Draw a diagram of the microscope and the telescope which you made, showing the positions of the lenses, the object, and image from the primary lens.

Conclusions

1. What do you observe about the images in each case? What happens when you move the secondary lens? the primary?
2. In each case, the secondary lens is making an image of the real image produced by the primary lens. What kind of image does the secondary lens form? What does your diagram tell you about this image?
3. Under what conditions did you get the best image from your lenses?

Quiz #40

Telescopes and Microscopes

Name: ______________________

Date: ______________________

1. The microscope and telescope discussed in this lab use how many lenses?
   a. 1
   b. 2
   c. 3
   d. 4

2. The primary lens in a microscope or a telescope forms a
   a. real image.
   b. virtual image.
   c. complex image.
   d. prime image.

3. You find the focal length of the lenses in this lab by
   a. focusing the light source.
   b. looking through the lens at an object.
   c. making an image of a distant object on a flat surface.
   d. doing none of the above.

4. With a converging lens, parallel rays of light are
   a. caused to diverge.
   b. caused to invert.
   c. focused at a point.
   d. enlarged.

5. The object used for the telescope is
   a. the light on the optical bench.
   b. your finger.
   c. the window.
   d. an object outside the window.

Lab #40: Telescopes and Microscopes

Teacher Notes

Cautions

1. Not being careful in your measurement of the focal lengths will cause problems.
2. Do not drop the lenses because they will break.

In this lab students learn how simple microscopes and telescopes are made. These two devices are important for other branches of science, mainly biology and astronomy. Working with these two devices should help students to realize that they are, in principle, rather easy to make. This lab should be done after the lab on lenses (#39). If the students are not familiar with the apparatus and how to find images from lenses, then this lab is difficult.

The lenses for this lab should be chosen carefully. A big lens (7.5-cm diameter or larger) is needed to gather enough light for the students to see the image. It should have a focal length of about 20 cm, and the small lens should have a focal length of about 5 cm. This will allow the images to be focused as well as possible. A few other lenses with focal lengths less than 20 cm are needed for the secondary lenses, or eyepieces, of the telescope.

Possible Problems

1. If the lenses are of very poor quality, the images formed will not be impressive.
2. The lenses do not gather enough light to give a good image. This is why bright objects and a large primary lens are important.

Conclusions

1. The images for both the microscope and the telescope should be enlarged and inverted. They are focused and unfocused as either of the lenses is moved. The enlargement remains the same as long as the focal length of the lenses is not changed.
2. The secondary lens creates a virtual image because the object (which is the image from the primary) is inside of the focus. The image is enlarged.
3. The quality of the image depends on the quality of the lenses and how they are placed.

(continued)

Lab Quiz Answers

1. b
2. a
3. c
4. c
5. d

Lab #41

Name: ______________________________

Date: ______________________________

A Sweet Lab on the Index of Refraction

In this lab you will investigate the change in the index of refraction of water when sugar is added. Sugar changes the properties of water and the index of refraction also changes. You will find the index of refraction for water and then slowly add sugar to it using the critical angle to observe how the index of refraction changes.

Materials

- tape
- protractor
- water container
- water
- paper towels
- light source
- sugar
- measuring device for the sugar
- stirring rod

Procedure

1. Tape a protractor to the bottom of the plastic, half-circle container. The reference point for measuring the angles should be at the center of the flat side of the container. Make sure the protractor is positioned correctly. Ask your teacher to check how you attached your protractor (see Figure 41).
2. Fill the container about two-thirds full with water, set it on a few paper towels, and set up the light source so that it shines through the water and hits the flat side. Make sure that the beam of light is as thin as possible.
3. Start with the beam of light striking the center of the flat side at zero degrees and increase the angle until you have total internal reflection. You can see if any light escapes by putting a piece of paper outside the container on the flat side of the container. Adjust the beam of light so that you have the smallest angle possible for total internal reflection. Precisely measure the angle which the light strikes the flat side and record it.
4. Add the suggested amount of sugar to the water and stir it until it dissolves. Again find the smallest angle for total internal reflection.
5. Repeat step 4 until sugar will no longer dissolve.

(continued)

Lab #41

A Sweet Lab on the Index of Refraction *(continued)*

Name: ______________________

Date: ______________________

Cautions

1. Make sure that you find the critical angle and not an angle larger than the critical angle. Larger angles will also cause total internal reflection.
2. Make sure that you do not get water on the light source.
3. Avoid spilling the sugar solution as it will make things very sticky.

Figure 41

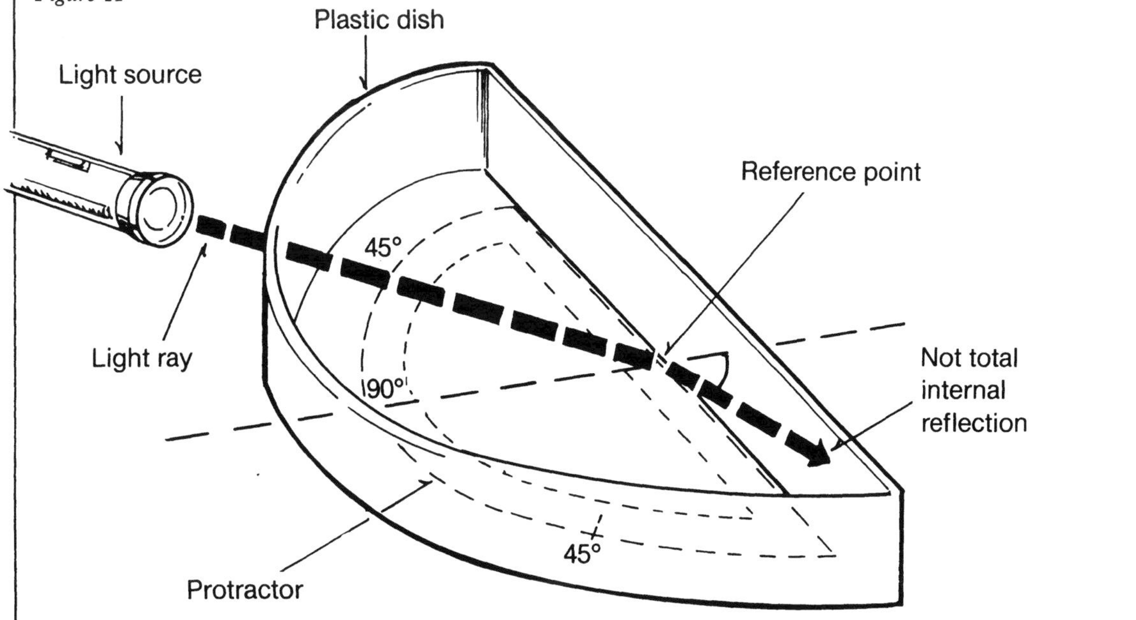

Analysis

1. Draw a graph of critical angle vs. amount of sugar added. Use the amount of sugar as the horizontal axis. On the vertical axis with the angle, start with your lowest angle right at the horizontal axis and use the entire vertical axis for amount which the angle changed. For instance, if the angle changed from 20 to 15 degrees, place 15 at the horizontal axis and divide the vertical axis into 5 units so that 20 is at the top of the page.

(continued)

Lab #41

A Sweet Lab on the Index of Refraction *(continued)*

Name: ______________________

Date: ______________________

2. The index of refraction is found using the equation for the index of refraction: $N = \sin(\theta^{air})/\sin(\theta^{medium})$. Find the index of refraction when no sugar was present, when half the sugar was present, and when all of the sugar was present.

Conclusions

1. What happens to the critical angle as more and more sugar is added? Give data to support your conclusion.
2. What happens to the index of refraction as more and more sugar is added? Give data to support your conclusion.
3. How are the critical angle and the index of refraction related? Explain.

Quiz #41

A Sweet Lab on the Index of Refraction

Name: ______________________________

Date: ______________________________

1. The speed of light in air divided by the speed of light in a medium approximates the __________________ for water.
   a. critical angle
   b. index of refraction
   c. angle of incidence
   d. total internal reflection

2. The smallest angle where total internal reflection occurs is called the
   a. critical angle.
   b. index of refraction.
   c. angle of incidence.
   d. angle of reflection.

3. The protractor is
   a. moved for each measurement.
   b. placed in the plastic dish.
   c. taped on top of the plastic dish.
   d. taped on the bottom of the plastic dish.

4. The beam of light which you shine on the flat edge of the plastic dish must
   a. hit the center of the flat side from the sugar water side.
   b. hit the center of the flat side from the air side.
   c. hit the end of the flat side from the sugar water side.
   d. hit the end of the flat side from the air side.

5. After you add sugar, the first thing that you must do is
   a. measure the angle of incidence.
   b. find the critical angle.
   c. stir the sugar until it is dissolved.
   d. none of the above.

Lab #41: A Sweet Lab on the Index of Refraction

Teacher Notes

Cautions

1. Make sure that you find the critical angle and not an angle larger than the critical angle. Larger angles will also cause total internal reflection.
2. Make sure that you do not get water on the light source.
3. Avoid spilling the sugar solution as it will make things very sticky.

In this lab students use the critical angle for the plotting of a graph of the index of refraction of sugar solutions. The most important concept is that of the critical angle. By working with the critical angle, students become more familiar with it and they gain understanding.

A plastic dish in the shape of a semicircle is required if the measurements are going to be made easily. Otherwise, it is hard to get the beam of light to meet the flat boundary at the proper place (i.e., the reference point on the protractor) and at the proper angle. This kind of dish is available from Central Scientific as a semicircular cell for about $4.

The light source must give a narrow beam of light which does not disperse and is bright enough for students to observe. A laser is ideal for this and could be used if you do the lab as a demonstration. A "Low-Cost Ray Box," which also includes a series of optics materials, is available from Science Kit for about $50. Many science departments have other light boxes which range from lights in tin cans, to flashlights with slits, to projector bulbs. A flashlight with a covering which has a small slit works and is relatively inexpensive.

Students must make sure that the ray of light hits the flat surface at the reference point and the reference point is right at the edge of the semicircular cell. Most protractors have a hole at the reference point and angles are measured from this hole, not from a point at the edge of the protractor.

Possible Problems

1. Students may not find the critical angle but find an angle somewhat larger.
2. Students may not stir the sugar completely and leave some undissolved at the bottom of the container.

(continued)

Teacher Notes (continued)

Conclusions

1. The critical angle gets smaller. It should change from about 50 degrees to about 42 degrees.
2. The index of refraction gets bigger. It should change from about 1.3 to 1.5.
3. The critical angle and index of refraction are inversely related. When one goes down, the other goes up.

LAB QUIZ ANSWERS

1. b
2. a
3. d
4. a
5. c

Lab #42

Radioactive Decay

Name: ______________________

Date: ______________________

In this lab you will investigate radioactive decay by using pennies to simulate decaying nuclei. In this lab, a penny will decay when it comes up heads (although you might not have known that this happened to pennies). You will investigate the half-life of pennies and you will take a look at probability with regard to small numbers and large numbers. The difference is important. When you throw 8 pennies, you would expect 4 pennies to come up heads the most often, but sometimes different numbers of heads will occur. You would actually be surprised if 4 pennies came up heads every time.

Materials

8 pennies or other coins

Procedure

1. You will throw 8 pennies for each decay cycle. A decay cycle for this lab will be until you have 1 or 0 pennies left. You will do 50 of these decay cycles. This sounds like a lot but each cycle can be done in much less than a minute even when taking the required data. Set up a data table to do this quickly. A sample is given below.

Data Table

Trial Number	Number Decayed 1st Time	Number of Throws to Get 1 or 0
1	6	3
2	3	4 etc.

2. For each trial, record the number of heads that shows up on the first throw. Also record the number of throws that is required to have one or no pennies left.

Analysis

1. Find the total number of pennies that "decayed" on the first throw by adding the "Number Decayed 1st Time."

(continued)

Name: ____________________________

Date: ____________________________

2. The total number of pennies you threw on the first throw is 50 trials times 8 pennies each trial. Find the percent of the total number of pennies thrown which "decayed" the first time.
3. Draw a bar graph of the number of pennies which decay on the first throw showing the number of trials when a given number of pennies "decayed." (If 2 decayed on the first throw 6 times, draw a bar up from the horizontal axis 6 units at the *2* position.)
4. Find the average number of throws required for 1 or 0 pennies to be left. This is the average value for three half-lives.
5. Draw a bar graph of the number of throws needed to get 1 or 0 pennies showing the number of trials when it took a given number of throws. (If twice it took 7 times to have only 1 or 0 left, then draw a bar 2 units long at 7.)

Conclusions

1. Is the total number of pennies which decayed on the first throw expected? Theoretically 50% would decay each throw; did this happen?
2. Comment on whether the expected number of heads came up on the first throw on a *particular* trial. Choose a trial at random, say trial 34, and see if the expected number of heads showed up. Did you expect this?
3. Comment on whether the first half-life of the pennies is expected. One would expect half the pennies to decay on the first throw; did this happen?
4. Is the time for three half-lives expected?
5. Discuss error, if any, in this lab.
6. In radioactive decay, you do not know which particular atom is going to decay, but you know that a certain percentage is going to decay in a given amount of time. Does this lab show this kind of decay?

Quiz #42
Radioactive Decay

Name: ____________________

Date: ____________________

1. In this lab on radioactive decay, what decays?
   a. atoms
   b. molecules
   c. masses
   d. pennies

2. As radioactive atoms decay, it is ____________ to know which will decay next.
   a. possible
   b. sometimes possible
   c. occasionally possible
   d. never possible

3. When you throw 8 pennies in the air, the most probable occurrence is to have ____________ pennies come up heads.
   a. 3
   b. 4
   c. 5
   d. 8

4. When you throw 8 pennies in the air, you will often experience
   a. more than 4 pennies coming up heads.
   b. less than 4 pennies coming up heads.
   c. 4 pennies coming up heads.
   d. all of the above.

5. Throwing pennies is like decaying atoms because
   a. you never know which penny will decay.
   b. it can't be determined which atom will decay.
   c. probability determines whether a penny or an atom decays.
   d. all of the above.

Lab #42: Radioactive Decay

Teacher Notes

This lab helps students to understand half-lives and also increases their familiarity with probability. The decay of pennies is easy for them to understand and after they have thrown pennies for the 50 trials, they know why you cannot tell when a particular event is going to happen. Of course, the data has zero percent error and the materials are very easy to procure. Washers can be used but pennies are less expensive. The equipment costs you almost nothing.

You do have to ensure that the pennies do not have two heads and that no one has weighted them, but other than that, there are no particular restrictions on how you obtain your lab equipment.

Because nuclear physics tends to be at the end of the school year, this kind of lab is useful because, in addition to being fun, it teaches some important physics.

Possible Problems

1. Students may be confused by finding the totals of the pennies decayed on the first throw and also the average number of throws for 1 or 0 pennies to be left. The tasks are simple but sometimes students need a push. To find the total number of pennies decayed on the first throw, you add all the numbers in the column under that heading. To find the average number of throws for 1 or 0 pennies to be left, you need to add the numbers in that column and divide by 50.
2. Students sometimes think that 50 trials is a lot. Actually, it can be done in less than 15 minutes if the lab group is fast.
3. Pennies that land on edge (it happens) can be blown over or pushed away from the student.

Conclusions

1. On the average, 200 total would have decayed on the first throw. But probability does not predict an exact answer, only an expected value.
2. On a particular trial, you do not know how many heads will show up. You might be interested if 0 or 8 show up, but it is not unreasonable.
3. On the average, half the pennies decayed on the first throw.
4. On the average, it took three throws to be left with 1 or 0 pennies.
5. Errors in counting pennies that have decayed should not occur.

(continued)

6. One never knows when a particular atom is going to decay and one never knows when a particular penny is going to decay. This lab shows that over time you can predict the results of a large number of random events but not on an atom-by-atom level.

LAB QUIZ ANSWERS

1. d
2. d
3. b
4. d
5. d

Lab #43
A Decay Series

Name: ______________________

Date: ______________________

In this lab you will investigate how much of a particular isotope remains as atoms decay from one isotope to the next in a decay series. When a nucleus decays, with an alpha or beta particle, it turns into another kind of atom. Many times, this atom is radioactive so it decays, too. The nucleus changes from one kind to another in a set path toward a stable nucleus. This path of changes is called a decay series. Some isotopes have longer half-lives than others so more atoms of that variety will exist than of the shorter half-life isotopes. For instance, if one isotope has a half-life of a minute and another an hour, more atoms would exist of the hour half-life isotope because they don't decay as fast. Pennies will be used for this lab and they will decay under specific conditions.

Materials

25 pennies | 4 pieces of paper

Procedure

1. Place the four pieces of paper on a table and label them *A*, *B*, *C*, and *D* going from left to right. Place the 25 pennies on *A*.
2. Set up a data table to record your data quickly. A sample is given below.

Data Table

Decay Cycle Number	Number in A	Number in B	Number in C	Number in D
0	25	0	0	0
1	12	13	0	0
2	5	14	6	0 etc.

Case 1:

3. Each decay will occur with the same half-life. A penny will decay if it comes up heads. For decay cycle 1, shake the pennies on *A* and move any which come up heads to *B*. They have decayed into *B*. In your data table, record the number of pennies on *A*, *B*, *C*, and *D*. (*C* and *D* will of course have no pennies.)
4. For decay cycle 2, start with the group of pennies farthest to the right (in this case those on *B*). Throw them and move any pennies which come up heads to the

(continued)

Name: ______________________

Date: ______________________

right onto *C*. Then throw the pennies on *A* and move any pennies which come up heads to the right to *B*. Record the number of pennies on each paper in the data table.

5. Repeat the process for decay cycle 3 but now there are pennies on *C* so you must throw them, the group farthest to the right, first. Throw *C* and move all heads to the right to *D*, throw *B* and move all heads to the right to *C*, etc. When each pile has been thrown, record the number of pennies in each pile.
6. Repeat the process for *eight* decay cycles. A decay cycle occurs when you shake the pennies in each group, going from right to left, and move the decayed pennies to the right.

Case 2:

7. Now the decay of pile *C* will be a different half-life. Only one quarter of the heads will decay. Count the number of heads which come up on *C* and divide by 4. Round to the nearest whole number and move that number of heads to pile *D*. Write on paper *C* "move one quarter of the heads." This will remind you that this is how they will decay.
8. Start with 25 pennies on *A* and start the decay cycles. Throw pennies until you have done *eight* decay cycles.

Case 3:

9. Now the decay of pile *B* will be a different half-life. Only one quarter of the heads will decay. Count the number of heads which come up on *B* and divide by 4. Round to the nearest whole number and move that number of heads to pile *C*. Write on paper *B* "move one quarter of the heads." Pile *C* pennies will decay very rapidly. Write "move all heads and half the tails" on paper *C*. You will move all heads and half the tails from *C* to *D*.
10. Start with 25 pennies on *A* and start the decay cycles. Throw pennies until you have done *eight* decay cycles.

Caution

Make sure that you move the pennies in the proper direction and that you have them decay according to the rules set down.

(continued)

Lab #43

A Decay Series *(continued)*

Name: ______________________

Date: ______________________

Analysis

1. Draw a graph with number of pennies on the vertical axis and number of decay cycles on the horizontal axis.
2. Plot the data from Case 1 on this graph. Use different symbols for each of the piles of pennies (*A*, *B*, *C*, and *D*) and draw the number of pennies for each of the four piles on the same set of axes. You may find it helpful to use different colors for each line.
3. Repeat steps 1 and 2 for Cases 2 and 3.

Conclusions

1. In Case 1, what happens to the number of pennies in pile *B* as time goes by? Do you expect this result? Explain.
2. In Case 1, how does the number of pennies in pile *D* change with time? This is the pile where no pennies decay.
3. In Case 2, what happens to the number of pennies as time goes by in pile *B*? in pile *C*? in pile *D*?
4. In Case 3, how does the number of pennies in pile *B* change with time? Do you expect this result? What happens to pile *C*?
5. Which decaying group has the longest half-life and which has the shortest? Explain.
6. You were trying to use radioisotopes to find out information about the history of rocks. You know that a particular atom decays from *A* to *B* to *C* to *D*, but you cannot find any of isotope *B* in the rock. What could explain this?
7. Lead-214 decays into Bismuth-214 in 27 minutes which decays in 20 minutes into Polonium-214 which decays in .0002 seconds into Lead-210 which takes 22 years to decay into Bismuth-210. What isotope would you expect to see the most of in a sample? the least? Explain.

Quiz #43
A Decay Series

Name: ______________________

Date: ______________________

1. In this lab on radioactive decay, what decays?
   a. atoms
   b. molecules
   c. masses
   d. pennies

2. Radioactive decay causes a nucleus to change from one kind of nucleus to another in what is called
   a. a decay change.
   b. a radioactive change.
   c. a decay series.
   d. a radioactive series.

3. As nuclei decay from one kind to another, some types of nuclei are more common than others because they
   a. are more important.
   b. have a larger size.
   c. have a shorter half-life.
   d. have a longer half-life.

4. Two isotopes of uranium differ in ______________ they have.
   a. the number of protons
   b. the number of neutrons
   c. the number of electrons
   d. the number of protons, neutrons, and electrons

5. Throwing pennies is like decaying atoms because
   a. you never know which penny will decay.
   b. it can't be determined which atom will decay.
   c. probability determines whether a penny or an atom decays.
   d. all of the above.

Lab #43: A Decay Series

Teacher Notes

Caution

Make sure that you move the pennies in the proper direction and that you have them decay according to the rules set down.

This lab helps students understand decay series and why particular isotopes are more common than others. It seems a bit complicated to run, but it actually is quite easy to throw the pennies and take the data. Each of the trials can be done in less than 10 minutes. You may want the students to repeat each case once to get a better idea of what is happening. Getting this extra data will not take twice as long because the students know what they are doing.

The equipment costs almost nothing and is readily available. Pennies cost less than washers and most of them will come back at the end of the lab. Students enjoy doing this although throwing pennies for more than 30 minutes will start to be boring.

You should try the experiment before you have students run it so that you can answer questions about it. Doing it yourself is actually quite fun, as well. Because nuclear physics tends to be at the end of the school year, this kind of penny-throwing lab may be useful because it is fun and teaches some important physics.

Possible Problems

1. Students may be confused by the instructions. It helps to demonstrate how the pennies decay before students start on the lab. Once they get going, they find that it is easy to gather the data.
2. Don't let students use pennies with two heads.

Conclusions

1. The number of pennies in pile *B* should increase and then decrease as the number of pennies in pile *A* is reduced.
2. Pile *D* has no pennies for a while and then steadily increases.
3. Pile *B* behaves the same way as Case 1. Pile *C* has zero for one decay cycle and then it increases rapidly to a maximum and then slowly becomes smaller. Pile *D*

(continued)

has zero for a while and then increases very slowly. The decay from pile *C* is very slow.

4. The number of pennies in pile *B* increases rapidly and then decreases slowly. The number of pennies in pile *C* never reaches a large number because pennies decay from pile *B* slowly and then they quickly decay from pile *C*. It fluctuates because of the random nature of decay but it is never very large as compared to the other cases.
5. The longest half-lives are pile *C* in Case 2, and pile *B* in Case 3. The shortest half-life is pile *C* in Case 3.
6. If the half-life of that isotope is very short, then few atoms of that isotope will be present.
7. Lead-210 will have the most atoms because it has a long half-life as compared to the others. Polonium-214 will have the fewest because it decays so quickly.

LAB QUIZ ANSWERS

1. d
2. c
3. d
4. b
5. d

Lab #44
Star Counting

Name: ______________________________

Date: ______________________________

In this lab you will investigate the number of stars which you can see with your naked eye. Of course, you need a night when the stars are visible and you are away from city lights, but with good "seeing" visibility, you can see thousands of stars. Obviously clouds or a bright moon will make it difficult to see stars. By using a simple technique, you can get an estimate of the number of stars you can see in the sky. You will not count each star but you will count a representative sample of them.

Materials

standard tube	small flashlight
meter stick	

Procedure

1. Measure the length and radius of your standard tube. Record these values.
2. On a clear night, go outside and let your eyes get adjusted to the dark by not looking at lights for about five minutes. (You will see many more stars after waiting than if you look immediately.)
3. Look through your standard tube at the stars. Turn in a random direction, hold the tube steady, and then count the stars which you can see through the tube. If you can't see any stars through the tube, record zero for this trial. Only if you point your tube at a cloud should you disregard that direction. At some times of year, it is possible to see the Milky Way which is our galaxy. The Milky Way is a band of light across the sky made up of millions of stars. You cannot see any of them individually so count only the individual stars which you can see.
4. Use the small flashlight to help you to write down the data. However, be careful not to put the brightest part of the beam on the white paper because it will cause you to lose your night vision.
5. Repeat the pointing of your tube in random directions 25 times. Each time, record the number of stars which you see. Make sure that you cover the entire sky. Count stars directly overhead and at all angles going to the horizon.
6. Record any observations you have about how clear the sky is. Can you see stars on the horizon? Have you seen the sky clearer? Were there any clouds? etc.

(continued)

Lab #44
Star Counting *(continued)*

Name: ______________________________

Date: ______________________________

Caution

Be careful that you take a random sample of the stars. Point the tube randomly but cover all parts of the sky.

Analysis

1. Using the radius, r, of your tube, find the area of the end of the tube. (Area = πr^2.)
2. Using the length of the tube, l, find the surface area of a sphere of that radius. (Area = $4\pi l^2$.) The area of the end of the tube is a part of the entire surface of the sphere.
3. Find how many times larger the surface area of the sphere is than the surface of the end of the tube. (Divide the answer in step 2 by the answer in step 1.)
4. Find the average number of stars which you saw each time you looked through the tube.
5. Multiply the average number of stars by the value you found in step 3.

Conclusions

1. Does the number of stars in the sky seem like a reasonable number? Is it larger or smaller than you thought? Explain.
2. In some directions, there are fewer stars than average. Why might this be? Explain.
3. In some directions, there were more stars than average. What might explain this?
4. In which direction or directions did you see the most stars? the least?
5. If you had a large telescope which could pick out individual stars in the Milky Way, how would you go about finding the number of stars in the Milky Way?

Quiz #44
Star Counting

Name: ______________________

Date: ______________________

1. In this lab, you are going to
   a. count all the stars in the sky.
   b. count half of the stars in the sky.
   c. count a sample of the stars in the sky.
   d. count stars in a constellation.

2. You need good "seeing" conditions to do this lab, meaning that
   a. you must be away from city lights.
   b. you must not have clouds in the sky.
   c. you must not have a bright moon in the sky.
   d. all of the above are necessary.

3. You cannot start counting stars right after you leave a bright room because
   a. your eyes are not night adjusted.
   b. you must find the direction where the most stars are before you start counting stars.
   c. you cannot count stars immediately after you go outside.
   d. stars do not come out until a few minutes after you go outside.

4. To count the stars, you point your tube
   a. where the most stars are.
   b. where the fewest stars are.
   c. only near the horizon.
   d. in random directions.

5. If when you point your tube no stars are visible, you should
   a. find a new direction.
   b. randomly point your tube in a new direction.
   c. move your tube until you see a star.
   d. record "zero" stars and then point your tube in a new direction unless you are looking at a cloud.

Lab #44: Star Counting

Teacher Notes

Caution

> Be careful that you take a random sample of the stars. Point the tube randomly but cover all parts of the sky.

This lab introduces students to the sky and gives them an idea of how many stars you actually can see. It focuses their attention on stars and makes them think about how some areas in the sky have more stars than others. It also makes them look at the "seeing" of the evening—how clear the night is and how easy it is to see stars. They will find that it is harder to see stars at the horizon than it is overhead because of more atmosphere in the way.

The "standard tube" can be made out of almost anything, but a durable tube which has about the right ratio of length to diameter is the tube on the inside of a toilet paper roll. This is an almost inexhaustible resource! Paper towel tubes can also be cut to the approximate dimensions. Tubes can also be made out of paper. A tube of these dimensions looks at about .5% of the sky. Students may often find that they can see no stars in the direction that they look. This is important for them to understand.

Students should do this lab keeping in mind that changes in the weather may affect what they can see. The timing of the lab may be difficult and it may not be possible to do this lab in the city at all. City lights or a bright moon (greater than half) make the number of visible stars very small and weather conditions may do the same. If the humidity is high, few stars can be seen. Fall and winter are preferable times to do the lab. In most parts of the country spring weather tends to be unpredictable and summer has high humidity.

In some places it may not be safe for students to be outside counting stars. Generally, these places are also near city lights so the lab cannot be done anyway. Do not encourage your students to take data if there is a chance of their being in danger when they are outside. You might recommend that the students only take data with a parent with them. This will involve the parent in the lab—an added benefit.

The lab can be used along with an exercise on learning a few constellations. The time required to get the data for this lab is not great and the calculations are easy. With the extra time, students could be asked to look at constellations which are visible at the time of year they do the experiment.

(continued)

Teacher Notes (continued)

Possible Problems

1. Weather conditions, a full moon, or city lights make the number of stars visible very small.
2. Students may not take a random sample of directions. The angles above 45 degrees tend to be underrepresented.

Conclusions

1. Under good conditions, one to two thousand stars can be seen with the naked eye.
2. Some areas of the sky have fewer stars than others. It is probability.
3. Some stars come in clusters so more stars are visible.
4. Most stars are seen overhead and the least at the horizon because of more atmosphere in the way when you look at the horizon.
5. To count the stars in the Milky Way, you would use the same general process of looking at small sections, finding the number of stars in those sections, and then multiplying by the number of similar sections which exist.

LAB QUIZ ANSWERS

1. c
2. d
3. a
4. d
5. d

Lab #45

Retrograde Motion

Name: ______________________

Date: ______________________

In this lab you will investigate retrograde motion and get a better understanding of why it occurs. Retrograde motion is the apparent backward motion as seen from the Earth when an inferior planet overtakes the Earth, or the Earth overtakes a superior planet (see Figure 45.1). (An inferior planet has a smaller orbital radius than the Earth while a superior planet has a larger orbital radius.) The planet does not change direction in its orbit; it is just that we see the motion from our moving reference frame. You will observe the motion of two planets as they orbit the Sun and find directions to the Sun and planet from the Earth.

Figure 45.1

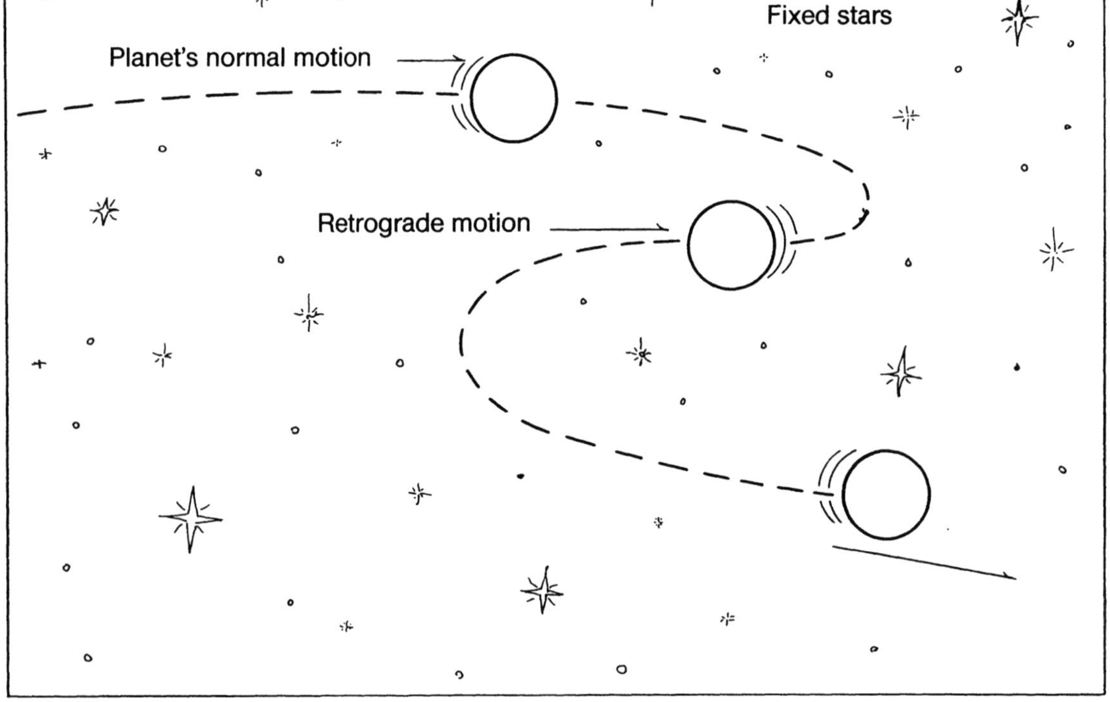

Materials

paper with orbits drawn on it

string

protractor

(continued)

Name: ______________________

Date: ______________________

Procedure

1. Choose whether you will use "Earth and Mars" or "Earth and Mercury." Then divide the circles into the number of divisions given below. Your divisions will indicate equal amounts of time.
2. If you are using Earth and Mars, divide the Earth orbit into 16 equal parts, and divide the Mars orbit into 32 equal parts. If you are using Earth and Mercury, divide the Mercury orbit into 8 equal parts, and divide the Earth orbit into 32 equal parts.
3. Make a data table that has a column for each of the following: "time interval," "angle to Sun," "angle from Earth to other planet," and the "angle between the Sun and your planet."
4. The direction to the top of the page will be your reference direction. Call it zero degrees. You will measure angles to that direction.
5. Start taking data with the Earth and your planet on the opposite side of the Sun from each other. The other planet should be near the top of the page and Earth toward the bottom. The Sun and planet should both be toward the top of the page or at zero degrees. If you are using Mercury, it should be on the inner orbit. If you are using Mars, it should be on the outer orbit.
6. Now number the divisions in each orbit going counterclockwise from where the planets are initially. Start with *0* and number them up to *32*. Some of the planets will have to go around more than once so put two or more numbers near each division if you have to.
7. Now start taking data. The time is initially *0* and the angles to the Sun and the planet are both *0*. Tie some string to the reference point on the protractor. You can then easily read the angles by holding the protractor on the Earth and moving the string to the Sun and the other planet. With the string taut, the angle can accurately be read. Record the time interval, the angle to the Sun, and the angle to the planet.
8. Move the Earth and planet one division. Find the angle to the Sun and the angle to the planet. Record these values in your data table. Measure the angle with respect to the fixed direction which is toward the top of the page.
9. Repeat the previous step until one planet has overtaken the other and *passed it* by eight time intervals. If you have to change the orientation of the protractor, make sure that the angles keep changing as they have been changing. If you have angles like 160, 170, 180, the next number will be about 190 rather than 10 or 170. If the angle becomes larger than 360 degrees, add 360 degrees to the value on the protractor. The angles should keep changing gradually, not go from 360 to 0.

(continued)

Lab #45

Retrograde Motion *(continued)*

Name: ______________________

Date: ______________________

Analysis

1. Find the "angle between the Sun and your planet" by subtracting the angle to the Sun from the angle to your planet for each time listed.
2. Draw three graphs with time on the horizontal axis and angle on the vertical. The graphs should be "time vs. angle to the Sun," "time vs. angle to the planet," and "time vs. angle between the planet and the Sun."

Conclusions

1. Comment on the shape of each of the curves.
2. Indicate when retrograde motion occurs. Why does it occur at this time? Explain.
3. Is the planet ever 180 degrees away from the Sun as viewed from Earth? Explain.
4. Explain why retrograde motion occurs based on your data.
5. Explain what each of the different graphs represents and what it would tell you about the motion of the Sun or planet in the sky.

Quiz #45
Retrograde Motion

Name: ______________________________

Date: ______________________________

1. Retrograde motion is
   a. backward motion of a planet in its orbit.
   b. apparent backward motion of a planet in its orbit.
   c. backward motion of the Earth in its orbit.
   d. apparent backward motion of the Earth in its orbit.

2. An inferior planet is one with
   a. a larger radius.
   b. a smaller radius.
   c. a larger orbital radius.
   d. a smaller orbital radius.

3. In this lab, measurements of ____________________ are taken repeatedly.
   a. mass
   b. distance
   c. time
   d. angle

4. When the Sun is in the direction of the top of the paper from the Earth, it would be at ____________________ degrrees.
   a. 0
   b. 90
   c. 180
   d. 270

5. You will draw graphs of
   a. angle vs. distance.
   b. distance vs. time.
   c. angle vs. time.
   d. velocity vs. time.`

Lab #45: Retrograde Motion

Teacher Notes

In this lab students visualize the motion of the Sun and planets. The lab is easy to set up and the results are convincing to students. The directions may seem to be a bit confusing at first, but once students get started taking data, things go very smoothly. It may help to show them on the board how to take the first few bits of data before you set them off on the lab. It will help them to take data quickly.

The materials for this lab are simple and the students could draw them. However, it is easier if you provide the circles which have been run off on a copier. The circles tend to be better drawn and of better sizes. It may even be helpful if you make the divisions on the circles to show the equal units of time. You must make one set of circles for Earth and Mercury and another for Earth and Mars.

The circles should be drawn concentrically. Mercury has an orbit 40% of the Earth's and Mars has an orbit 50% larger than the Earth's. You may wish to have the students keep these circles for their lab write-ups or they may leave them behind for others to use. Figures 45.2 and 45.3 can be copied for this purpose.

Possible Problems

1. If the students do not keep measuring relative to the same direction, the top of the page, then their data will be incorrect. Make sure that students understand this.
2. Students sometimes run into trouble with measurement when they go past 180 degrees. Then they have to turn the protractor around and this sometimes confuses them.

Conclusions

1. The angle to the Sun increases at a constant rate as the Earth goes around the Sun. The angle to the planet generally increases except when retrograde motion occurs and the angle becomes less. The angle between the planet and the Sun is different for the superior planet, Mars, because Mars can be opposite or 180 degrees from the Sun. The inferior planet, Mercury, cannot be more than about 40 degrees away from the Sun, first on one side, then on the other. When it goes between the Earth and the Sun, the angle changes more rapidly than when it is behind the Sun.
2. Retrograde motion occurs when the angle to the planet decreases because one planet passes the other.
3. Only Mars gets 180 degrees from the Sun. Mercury never gets that far because it is an inferior planet.

(continued)

Figure 45.2

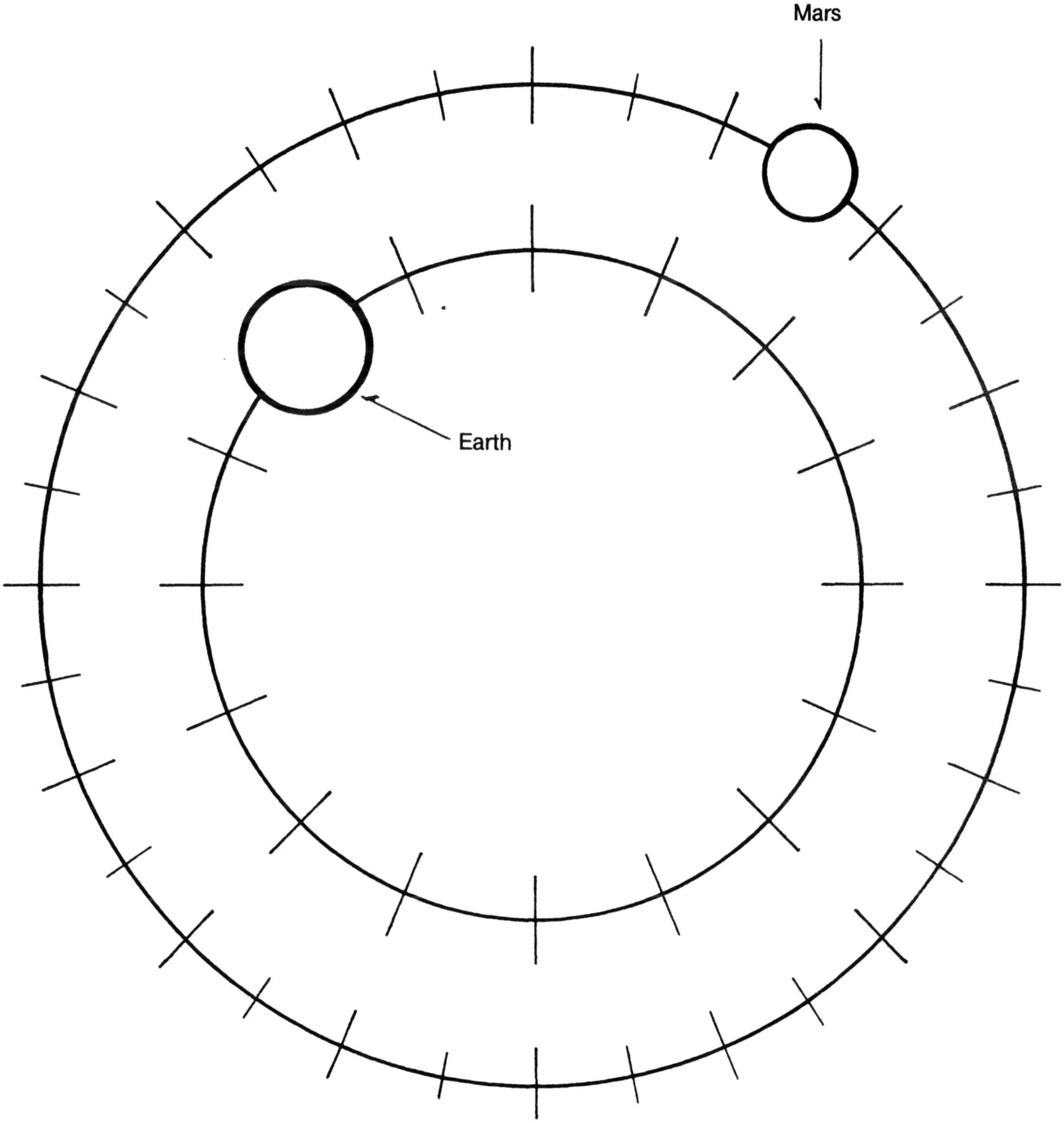

(continued)

Figure 45.3

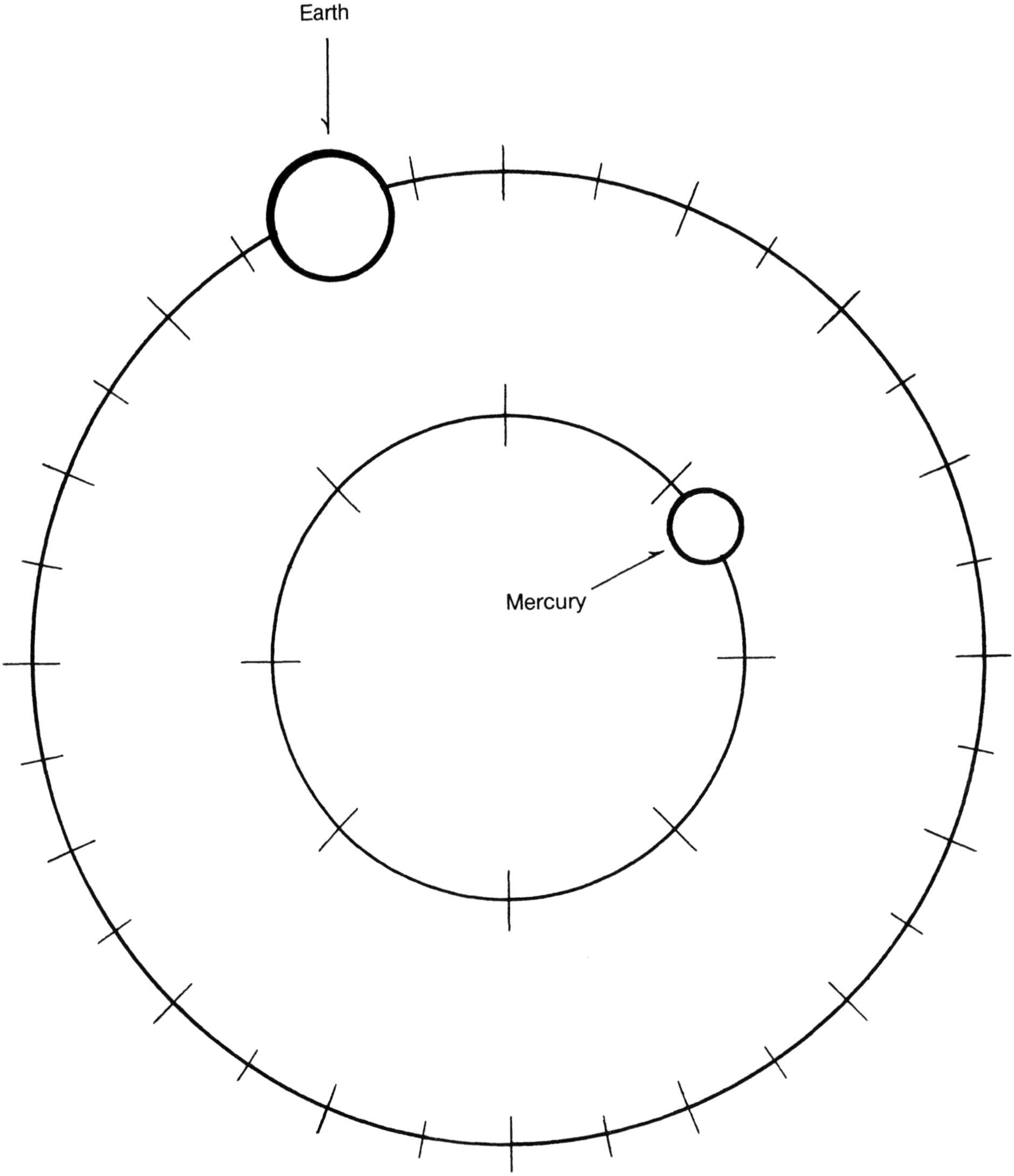

(continued)

Teacher Notes (continued)

4. Retrograde motion occurs when one planet overtakes the other. When retrograde motion occurs, the inner planet is between the outer planet and the Sun.
5. The first graph shows that the Sun moves at a constant rate across the sky, making one complete circle a year. The second graph shows how retrograde motion occurs. The third graph shows where to look for the planet relative to the Sun at different points in its orbit. Mars is sometimes up at midnight (180 degrees from the Sun) while Mercury never is.

LAB QUIZ ANSWERS

1. b
2. d
3. d
4. a
5. c